International Union of Crystallography
Commission on Crystallographic Teaching

CRYSTALLOGRAPHIC BOOK LIST

Edited by

HELEN D. MEGAW

(*Cavendish Laboratory, Cambridge, England*)

with the assistance of

H. CURIEN (*Laboratoire de Minéralogie-Cristallographie à la Sorbonne, Paris, France*)

E. G. STEWARD (*Northampton College of Advanced Technology, London, England*)

M. M. UMANSKIJ (*Moscow State University, U.S.S.R.*)

J. ZEMANN (*Mineralogisch-Kristallographisches Institut der Universität, Göttingen, Germany*)

March 1965

Copies of this Book List have been distributed free of charge to all subscribers to ACTA CRYSTALLOGRAPHICA, a journal of the International Union of Crystallography. Additional copies can be obtained from N.V.A. Oosthoek's Uitgevers Mij, Domstraat 11-13, Utrecht, The Netherlands, at the price of 10 Netherlands Guilders (U.S.$3.00 or U.K. £1 at the present rates of exchange) per copy, including postage for shipment by surface mail. In the event of foreign exchange difficulties, UNESCO coupons will be accepted. Orders can also be placed with Polycrystal Book Service, G.P.O. Box 620, Brooklyn 1, N.Y., U.S.A.; or with any bookseller.

CRYSTALLOGRAPHIC BOOK LIST

CONTENTS

GENERAL PREFACE

This is a Crystallographic Book List - but what is Crystallography? We take the following working definition.

> Crystallography is the branch of science concerned with the description and understanding of the structure and properties of condensed states of matter in terms of the spatial relationships of atoms and interatomic forces in an extended array.

Moreover

> A Crystallographer is a scientist with an active interest in Crystallography, either for its own sake or for the contribution it can make to some other branch of science. (Obviously this definition is not an exclusive one)

The problem of deciding what books should be included in this List made it necessary to think what topics are comprised in Crystallography, and to clarify the issue a draft syllabus was constructed, and is set out below. Though it probably covers too wide a spectrum for any one individual (or University Department) to deal with in detail, there is no part of it which would not be regarded by some Crystallographers as quite essential. It is apparent that some topics are central to Crystallography - notably crystal geometry and symmetry, diffraction by perfectly and imperfectly periodic structures, methods of structure analysis, and the relation of physical and chemical properties to structure - and others are peripheral, linking the central topics to other fields of knowledge. The peripheral topics are of great interest and importance - perhaps even the main interest of some Crystallographers - but where, in a Book List, does one draw the line?

It must be admitted that no very satisfactory and consistent solution has been reached, since no one person can be equally at home in all the fields or have read all the books, and standards of judgment vary not only from person to person but also for the same person from time to time. Very often the treatment and emphasis of the book, rather than its nominal subject matter, should decide the issue, and these cannot easily be assessed at second hand. Some factors favouring the inclusion of a borderline book are as follows: if the author is known to be a Crystallographer (for example, by inclusion in the World List); if a reviewer in Acta Crystallographica thinks it of interest to Crystallographers; if it has been personally recommended by a Crystallographer; if it represents a new approach; if it seems helpful in bringing crystallographic ideas into the teaching of related subjects. Reports of Conferences and collections of articles may be included for the sake of a proportion of relevant papers, even though others are irrelevant. Some books of only marginal interest may however occur because, once provisionally listed, inertia keeps them there unless further information comes to hand.

Only books written in one of the four official languages of the International Union of Crystallography - English, French, German, and Russian - have been included, unless either they are originals of books translated into one of the four languages or they have been reviewed in Acta Crystallographica. For books in French, German or Russian, I have relied on the advice of Professor H. Curien, Professor J. Zemann, and Professor M.M. Umanskij respectively, and I owe them very considerable thanks both for sending me lists of relevant books and for correcting and modifying the cumulated lists in draft form.

There is no restriction based on whether a book is still in print, and no clear-cut restriction by date, though broadly speaking all the older peripheral books have been left out unless they obviously influenced crystallographic thinking. From about 1950 onwards the list is a comprehensive and not a selected one (except at the periphery). Before that date there is some omission of the more ephemeral; and before 1935 the intention is to include only those of particular lasting interest. In the early range, however, the selection is likely to be uneven and arbitrary, as it depended on suggestions from correspondents and an inspection of the most accessible library shelves rather than on a proper bibliographical search.

All books published in 1964 for which adequate information was available in time are included in Lists I-IV; there is however a very variable time-lag between publication and the date when the Editor gets complete information, which means that some less recent books may be omitted while more recent ones are included. List V contains entries for which information came in after the other lists had gone to press, up to the final closing date in January 1965. A blank page follows it, where readers can make their own additions, and where they can insert short supplementary lists if (as is hoped) these are published at intervals in loose-leaf form.

It has not been possible to list individual papers or review articles, however important, if they were published as only part of a volume; nor to include privately-printed theses (unless inadvertently). To have done so would have meant much bibliographical research.

Journals appearing regularly in parts are not listed at all; collections of articles coming out at longer intervals as bound volumes are included, if their contents are relevant, and so are special volumes of Journals devoted wholly to Conference reports. Here too it is difficult to draw a consistent line.

The underlying aim of this Book List is to make it as easy as possible for users to trace the books, and the detailed conventions, described in the Preface to List I, have been devised with that in mind. Bibliographical accuracy of detail has not been an aim; indeed, it has been deliberately rejected at points where it would have made confusing complications or taken up time and space without giving real help in tracing the book - for example, in insisting on the form of the publisher's name recorded on the title page, or on consistency in the way the number of pages is counted.

No list of this kind can be perfectly free from mistakes. The Editor asks forgiveness for mistakes and omissions, and invites users to send in corrections, which will be collected and published when opportunity offers.

ACKNOWLEDGEMENTS

A debt of gratitude is owed to the Editors of the forerunners of this Book List, compiled for the International Union of Crystallography by its Commission on Crystallographic Teaching: namely, Dr. H.J. Milledge and Dr. A. Magneli, whose lists were respectively the starting point and the broadened foundation of the present work. As the work proceeded, many colleagues in different countries have helped by sending in lists or individual items; their help is gratefully acknowledged.

HELEN D. MEGAW
January 1965

A SYLLABUS FOR CRYSTALLOGRAPHY

This syllabus is put forward as a tentative draft, to help discussion, d not as in any way definitive. It may serve as a guide to those un- miliar with crystallography, or familiar only with limited aspects of it.

It is taken for granted that the principles and concepts listed will be veloped for, illustrated by, and applied to, the widest possible range of terials - including those of interest in metallurgy, "solid-state" and semi- nductor physics, "materials science", mineralogy, biochemistry and molecular ology, as well as the whole range of inorganic and organic chemistry - so at false generalisations based on too narrow experience are avoided and a od over-all perspective is achieved.

The left-hand column gives the central contents of crystallography. e right-hand column gives the other subjects needed as tool or background, least at a basic elementary level.

i) Geometry and symmetry

Geometry of solids. Point groups, space groups, lattices, periodic structures with atoms in general positions. Formal descriptions and methods of calculation	Solid geometry, spherical trigonometry (and the stereographic projection)
Colour groups	Formal group theory

ii) Diffraction theory and the concept of reciprocal space

The crystal as 3-dimensional grating; the Laue conditions; Bragg's law.	Diffraction of light
The reciprocal lattice.	
Physics of diffraction; coherent and incoherent scattering; dynamical theory; Mossbauer effect.	Production and properties of X-rays
Phase change on scattering.	
Imperfect crystals: thermal vibrations, substitution disorder, position disorder, faults, mosaic blocks, finite size (all treated theoretically)	

ii) Structure analysis by Fourier and "direct" methods*

Electron density as transform of structure amplitudes. Fourier maps (electron-density, Patterson, difference, generalised-projection, minimum-function, etc.)	Mathematical introduction to Fourier methods
Transforms of non-periodic units; optical transforms.	
The phase problem. Empirical methods, e.g. heavy-atom. "Direct" methods.	
Refinement techniques, including least-squares.	
Assessment of accuracy	Statistics

v) Diffraction techniques

X-ray powder: photographic and counter; high and low angles. Single crystal methods.	Production and properties of X-rays; absorption; filters
Monochromators. Measurements of intensity and spacing. Background scattering.	Physics of counters 3-dimensional geometry
Accuracy.	Error estimation
Neutron diffraction; electron diffraction; advantages and disadvantages.	Production and properties of neutrons

Syllabus for crystallography

(iv) continued

Applications to identification, solid solutions, phase changes, physical properties, nature and degree of imperfections

(v) Crystal chemistry

Geometry of actual structures. Concepts of atomic size, coordination, directed bonds.	The periodic table. Chemica treatments of valency
Packing structures and framework structures; topology of networks.	
Interatomic forces: simple classification and modern developments. The hydrogen bond.	Structure of the atom: s,p,d orbital and methods of combin tion
Molecules and heterodesmic structures	Stereochemistry of organic mo cules Molecular orbitals
Energies. Born treatment. Madelung constant	Physical chemistry of solids.
Bond energies. Crystal field theory. Ligand field theory. Magnetochemistry	Quadrupole resonance and nucl magnetic resonance.

(vi) Crystal physics*

Macroscopic physics of anisotropic materials; Symmetry of physical properties.	Physical concepts concerning heat and properties of matter for isotropic solids Vectors, tensors, and matrice
Explanation in terms of interatomic forces (qualitative or semi-quantitative)	
Lattice dynamics: Born approach; frequency spectrum; elastic constants; energy relations in terms of interatomic forces and lattic vibrations.	Wave propagation in isotropic solids; dependence on elasti constants Einstein and Debye theories o specific heat
Electron theory of solids; band theory; Brillouin zones; Fermi surface; defects, vacancies and interstitials	Conduction and semi-conductor
Dielectrics: piezoelectricity, pyroelectricity, ferroelectricity	
Ferromagnetism and antiferromagnetism; magnetic transitions	Magnetic domain textures
Optics. Double refraction; indicatrix; wave surface. Polarising microscope and its use. Rotary polarisation and absolute configuration. Spectra (u.v., optical, i.r. & Raman) in relation to lattice dynamics	Refraction in isotropic mater Polarisation of light Electromagnetic theory of lig Spectroscopy

(vii) Imperfections, morphology and growth*

Point defects Dislocations	Electron microscopy X-ray microscopy
Mistakes, stacking faults, antiphase boundaries, polytypes, "OD" structures	Field emission microscopy
Substitution disorder; nature of solid solution	
Twinning and twin laws (macroscopic and structural); polysynthetic twinning	
Morphology in relation to structure; epitaxy and pseudomorphy	
Nucleation and grain growth	Textures of polycrystalline materials
Surfaces and surface effects Energy relations: strain energy, domain walls, cold working.	Optical interferometry Strength of materials
Radiation damage; partly crystalline materials; glasses	

(viii) Transitions and changes of state*

Polymorphism	
Displacive and reconstructive transitions	
Martensitic and "diffusionless" transitions	
Order-disorder transitions	Statistical thermodynamics
Pseudosymmetric transitions and "superlattices"	
Mechanism of oriented transitions; character and role of diffusion; "unmixing".	
Effects of temperature, pressure, electric field, composition, etc.	
Effects of surface, size, and nucleation	Catalysis
Liquid crystals; melting.	Theory of liquids

* All these topics to be discussed for as wide a range of materials as possible

TRANSLITERATION OF RUSSIAN

а	a	и	i	р	r	ш	š
б	b	й	j	с	s	щ	šč
в	v	к	k	т	t	ы	y
г	g	л	l	у	u	ъ	ʼʼ
д	d	м	m	ф	f	ь	ʼ
е	e	н	n	х	kh	э	ė
ж	ž	о	o	ц	c	ю	ju
з	z	п	p	ч	č	я	ja

NOTES ON ABBREVIATIONS

(a) **Abbreviations other than publishers' names**

Abbreviations in languages other than English are distinguished as follows:

* German
** Russian

*	Bd.	Band [Volume]
	corr.	corrected
**	č.	čast' [Part]
	ed(s).	editor(s), edited (by)
	edn.	edition
**	Izdat.	Izdatel' stvo [Publishing house]
	No.	number
	Pt.	Part
	publ.	published, publisher(s)
	ref.	reference
	repr.	reprinted
	rev.	revised (by)
	Suppt.	Supplement
	tr.	translated (by)
	Univ.	University [or the corresponding word in other languages]
*	Verlagsg.	Verlagsgesellschaft
	Vol.	Volume

(b) Abbreviations of publishers' names

In general, the distinctive part of the name has been retained, whil first-name initials and descriptions such as "& Co. Ltd.", or "Uitgeversmaatschappij" are cut out. The words "Press", "Editorial", "Verlag", "Izdat(el'stvo)" are sometimes included and sometimes cut out.

Publishers commonly known by their initials or other abbreviation are listed below according to their countries.

FRANCE

C.N.R.S.	Centre Nationale des Recherches Scientifiques
P.U.F.	Presses Universitaires de France

SPAIN

C.S.I.C.	Consejo Superior de Investigaciones Cientificas

UNION OF SOVIET SOCIALIST REPUBLICS

(Much of the information in this list was supplied by Professor M.M. Umanskij, whose he gratefully acknowledged. Some of the names and abbreviations have been altered recent the new form, with alternatives if any, is given first, followed by older forms in brac and the full name, transliterated.)

(Akad.Nauk SSSR), (AN) -	[see NAUKA]
Akad.Nauk -SSR, Akad.Nauk SSR-	Akademii Nauk [of a particular Soviet Socialist Republic]
GNTI, ONTI, GONTI	Gosudarstvennoe naučno-tekhničeskoe izdatel'stv Gosudarstvennoe ob"edinenie naučno-tekhničeskik izdatel'stv
Gostekhizdat, Gostekhteorizdat	Gosudarstvennoe izdatel'stvo tekhniko-teoretiče literatury
Izdat. Univ.	[University press]
Khimija, (Goskhimizdat)	Gosudarstvennoe naučno-tekhničeskoe izdatel'stv khimičeskoj literatury
Mašinostroenie (Mašgiz)	Gosudarstvennoe naučno-tekhničeskoe izdatel'stv mašinostroitel'noj literatury
Metallurgija (Metallurgizdat)	Gosudarstvennoe naučno-tekhničeskoe izdatel'stv literatury po černoj i cvetnoj metallurgii
MIR (I.L.)	(Izdatel'stvo inostrannoj literatury)
NAUKA	(Izdatel'stvo Akademii Nauk SSSR)
NAUKA (Fizmatgiz)	(Gosudarstvennoe izdatel'stvo fiziko-matematiče literatury)
NEDRA (Gosgeolizdat)(Geolgiz)	(Gosudarstvennoe izdatel'stvo geologičeskoj literatury)
NEDRA (Gosgeoltekhizdat)	(Gosudarstvennoe naučno-tekhničeskoe izdatel'st po geologii, geodezii i okhrane nedr)
NEDRA (Gosgortekhizdat)	Gosudarstvennoe naučno-tekhničeskoe izdatel'stv literatury po gornomu delu
Prosveščenie, Vysšaja škola (Učpedgiz), (Učpedizdat)	Gosudarstvennoe izdatel'stvo učebno-pedagogičes

UNITED KINGDOM

H.M.S.O.	Her Majesty's Stationery Office

UNITED STATES OF AMERICA

A.C.A.	American Crystallographic Association
ASTM	American Society for Testing and Materials
ASXRED	American Society for X-Ray and Electron Diffrac - now incorporated in A.C.A.)

MAIN LIST - I

Explanation of conventions

Arrangement is alphabetical by author or editor (or sponsoring Society if no author or editor is named). Works by the same author are in chronological order, except that translations follow immediately after the work translated. Works normally remembered by their title and not their editor should be looked for in List III, unless they are Conference reports, to be found under their year in List II. Cross-references from this List are however given.

Alphabetical conventions

The order of the English alphabet is followed, with certain modifications as follows.

(i) Letters not in the English alphabet are alphabeticized by ignoring French accents, treating German ä, ö, ü, as if they were spelt ae, oe, ue, and in transliterations from Russian placing modified letters immediately after the corresponding unmodified letters, (e.g. č after c, š after s).

(ii) Names (of British origin) beginning with Mc or Mac, whether or not a capital letter follows, are all listed as if they began with Mac.

(iii) For prefixes such as de, van, von, the usage of the country where the author works is followed - in practice, the prefix is left out of alphabetical consideration for the Continent of Europe, but included for English-speaking countries.

(iv) Russian names have been transliterated by the convention used in STRUCTURE REPORTS, unless some other spelling is already in current use, in which case a cross-reference is given from the conventional form. Where different transliterations of the same name are current, a standard form is here chosen and is written first, but enclosed in square brackets if it is not the form used on the actual book, which then follows it, e.g. [Ioffe,A.F.=] Joffe,A.F., a cross-reference being given from Joffe.

Editors and revisers

Where a new edition, or a translation, has been revised by a second author, no consistent practice exists as to whether the reviser's name is given as co-author, or is added in brackets, or is omitted. Very occasionally it is put first, with a cross-reference from the original author.

Title

Some books have a short title on the cover and a longer form on the title page. The present list (like its sources) is inconsistent as to which form is used, or whether, in the long form, the extra words are put in brackets; but it is hoped that this will not lead to any ambiguities in identification.

The titles of Russian books are transliterated, and a translation is added [in square brackets]. Where necessary and possible, literal translations are replaced by the technical English phrases for the subject matter. Where the book is a translation into Russian, the title in the original language is given instead of a back-translation.

Where anything not in the title is added to help in identification, square brackets are used.

Publisher and date

The city of publication is given first, in its own language; it is followed (in brackets) by its country, or its State of the U.S.A., if confusion might arise: e.g. Cambridge (England) or Cambridge (Mass.).

The name of the publisher is given, where possible, in a short but distinctive standard form, irrespective of the form used on the title page. Abbreviations of publishers' names are listed on p. xii.

The dates of the first and the most recent edition are given, and sometimes also dates of recent reprintings (with or without revision) if there has been a long gap. Publishers' numberings have been followed, though they vary a good deal in meaning; some number every reprinting, and some only use a new number after substantial changes. No attempt has been made here to distinguish which new editions are revised editions; probably the majority of them are. Mention of revision of a text before translation is also sometimes omitted.

Apparent discrepancies may appear for the following reasons (which generally cause no confusion for experienced booksellers):
(i) Some publishing houses publish in more than one city; the city mentioned may be the actual place of publication or that of the publisher's present head office.
(ii) Some publishers give separate names to their subsidiary firms, or have sharing arrangements by which they re-publish the works of other firms under their own name; or a Scientific Society which is the original publisher of a book may later entrust it to a commercial publisher.
(iii) The year of publication may be wrong by ±1, because some sources quote the date on the title page, and some the actual date of appearance, which may be later.

Additional information

The first column of figures gives the approximate number of pages (intended only as a guide to the length of the book, and not as a detailed bibliographical check).

The second column gives a reference to the volume and page of *Acta Crystallographica* in which a review of the book has appeared.

The third column gives a subject classification number, or numbers, according to the main headings of List IV. (For a key to the numbers, see the preface to List IV.).

The final column gives an indication of the teaching level of the book, as assessed by a teacher using the language of the book. No attempt has been made to judge its merits. The fact that a book is not classed does not necessarily mean that it is unsuitable for teaching; it may just happen that no correspondent has happened to mention it to the editors. On the other hand, advanced treatises, specialist works and books of reference are deliberately not classed, and collections of articles are only classed if they contain something particularly suitable for teaching purposes. The classification is necessarily very rough as well as incomplete, but may be of use as a guide.

Key to code indicating teaching level

E - Elementary, pre-University, popular

U - University, undergraduate, (early years and main course)

G - Graduate, final University year(s), advanced course

Where two letters are used in combination, it means that parts of the book may be used at an earlier stage than the whole.

Main List - I

Author	Title, publisher, date	No. of pages	Review ref.	Subject ref.	Teaching level
Addison,W.E.	Structural principles in inorganic compounds -London: Longmans, 1961	183		7	U
Ageev,N.V.	Rentgenografija metallov i splavov [Röntgenography of metals and alloys] -Leningrad: KUBUCH, 1932	192		5,12	
American Society for Testing and Materials - see ASTM					
Amoros,J.L.	Cristaloquimica -Madrid: C.S.I.C., 1951	130	6,112	7	
Amoros,J.L.	Tecnica del analisis cristaloquimica -Madrid: C.S.I.C., 1952	130	6,368	5	
Amoros,J.L.	Cristalofisica Pt.1 -Madrid: Aguilar, 1958	233	11,702	9	
Andrew,E.R.	Nuclear magnetic resonance -Cambridge (England): Univ.Press, 1955	267		19	
Anšeles,O.M.	Vyčislitel'nye i grafičeskie metody kristallografii [Numerical and graphical methods in crystallography] -Leningrad: Izdat.Univ., 1939	300		2	
Anšeles,O.M.	Načala kristallografii [Principles of crystallography] -Leningrad: Izdat.Univ., 1952	276		1	
Anšeles,O.M. & Burakova,T.N.	Mikrokhimičeskij analiz na osnove kristallooptiki [Microchemical analysis based on crystal optics] -Leningrad: Izdat.Univ., 1948	133		14	
Arkel,A.E.van	Moleculen en kristallen -Den Haag: Van Stockum (1)1941, 1961	650		7	U
Arkel,A.E.van	(tr. from Dutch) Molecules and crystals in inorganic chemistry -London: Butterworths, (1)1949, (2)1956	270		7	U
Arkharov,V.I.	Kristallografija zakalki stali [The crystallography of tempering of steel] -Sverdlovsk-Moskva: Metallurgija, 1951	144		12,13	
ASTM	Properties of crystalline solids - see Conference 1960/1, List II				
ASTM	Advances in techniques in electron metallography - see Conference 1962, List II				
Azaroff,L.V.	Introduction to solids -New York: McGraw Hill, 1960	460		9	UG
Azaroff,L.V. & Brophy,J.J.	Electronic processes in materials -New York: McGraw Hill, 1963	462		9	

Main List - I

Author	Title, publisher, date	No. of pages	Review ref.	Subject ref.	Tea in lev
Azaroff,L.V. & Buerger,M.J.	The powder method in X-ray crystallography -New York: McGraw Hill, 1958	342	11,753	5	
Azaroff,L.V. & Buerger,M.J.	(tr. Akimov,V.M., from English, 1958) Metod poroška v rentgenografii [The powder method in X-ray crystallography] -Moskva: MIR, 1961	363		5	
Bacon,G.E.	Neutron diffraction -Oxford: Univ.Press, (1)1955, (2)1962	426	17,622	5	
Bacon,G.E.	(tr. Burštejn,E.L., from English, 1955) Difrakcija nejtronov [Neutron diffraction] -Moskva: MIR, 1957	256		5	
Bacon,G.E.	Applications of neutron diffraction in chemistry -Oxford: Pergamon, 1963	141		5,7	
Bagarjackij,Ju.A.	(ed.) Rentgenografija v fizičeskom metallovedenii [X-ray methods in physical metallurgy] -Moskva: Metallurgija, 1961	368		12,5	
Barber,N.F.	Experimental correlograms and Fourier transforms -London: Pergamon, 1961	136		3,6	
Barker,J.A.	Lattice theories of the liquid state -Oxford: Pergamon, 1963	133		18	
Barker,T.V.	Graphic and tabular methods in crystallography -London: Murby, 1922 [Still in print]	192		2	U
Barker,T.V.	The study of crystals: a general introduction -London: Murby, 1930	133		1	E
Barker,T.V.	Systematic crystallography -London: Murby, 1930	115		2	
Barker Index	- see List III				
Barraud,J.	Principes de radiocristallographie -Paris: Masson, 1960	236		1	
Barrett,C.S.	Structure of metals -New York: McGraw Hill, (1)1943, (2)1952	661	6,672	12,5	
Barrett,C.S.	(tr. from English) Structure des métaux -Paris: Dunod, 1957	618	11,452	12,5	
Barrett,C.S.	(tr. & ed. Umanskij,Ja.S. from English, 1943) Struktura metallov: kristallografičeskie metody, principy i dannye [Structure of metals] -Moskva: Metallurgija, 1948	676		12,5	
Bauer,E.	Elektronenbeugung -München: Verlag Moderne Industrie, 1958	240	14,442	5	

Main List - I

Author	Title, publisher, date	No. of pages	Review ref.	Subject ref.	Teaching level
Beeching,R.	Electron diffraction -London: Methuen, (1)1936, (3)1950	106		5	EU
Belaiew,N.T.	Crystallisation of metals -London: Univ.Press, 1922	143		1,12	E
Beljankin,D.S. & Petrov,V.P.	Kristalloptika [Crystal optics] -Moskva: NEDRA (1)1928, (4)1951	128		14	
Beljankin,D.S. & [Petrov,V.P.=]Petrow,W.P.	(tr. from Russian, 1951) Kristalloptik -Berlin: Verlag Technik, 1954	180		14	U
Belov,K.P.	Magnitnye prevraščenija [Magnetic transitions] -Moskva: NAUKA, 1959	259		9,16	
Belov,K.P.	(tr. W.H.Furrey from Russian, 1959) Magnetic transitions -New York: Consultants Bureau, 1962	242		9,16	
Belov,N.V.	Struktura ionnykh kristallov i metalličeskikh faz [Structure of ionic crystals and the metallic phase] -Moskva: NAUKA, 1947	237		7,12	
Belov,N.V.	Strukturnaja kristallografija [Structural crystallography] -Moskva: NAUKA, 1951	86	5,858	1,7	
Belov,N.V.	Klassnyj metod vyvoda prostranstvennykh grupp simmetrii [A class-room method for the derivation of space groups] -Moskva: NAUKA, 1951	62		2	U
Belov,N.V.	(tr. Balashov,V. from Russian, 1951) A class-room method for the derivation of the 230 space groups -Leeds: Philosophical and Literary Society [sold through Austick's Bookshop, Leeds 2], 1957	46	14,899	2	U
Belov,N.V.	Kristallokhimija silikatov s krupnymi kationami [Crystal chemistry of silicates with large cations] -Moskva: NAUKA, 1961	66		8,15	
Belov,N.V. (and others)	(tr. from Russian, 1961) Crystal chemistry of large-cation silicates -New York: Consultants Bureau, 1963	168		8,15	
Belov,N.V. (and others)	(tr. from Russian) Coloured symmetry - see Shubnikov & Belov				
Bentley,W.A. & Humphreys,W.L.	Snow crystals -New York: McGraw Hill, 1931 -New York: Dover, repr. 1962	226		13,21	

Main List - I

Author	Title, publisher, date	No. of pages	Review ref.	Subject ref.	Teac ing leve
Berek	- see Rinne & Berek				
Bergmann,L.	Der Ultraschall und seine Anwendungen in Wissenschaft und Technik -Stuttgart: Hirzel, (2)1939, (6)1954	1114	5,298 8,69	9	G
Bergmann,L.	Der Ultraschall: Nachtrag zum Literaturverzeichnis der 1954 Erscheinung -Stuttgart: Hirzel, 1957	68	11,60	9	G
Bergmann,L.	(tr. Hatfield,H.S. from German) Ultrasonics and their scientific and technical applications -New York: Wiley, 1948	264		9	
Bergmann,L.	Schwingende Kristalle und ihre Anwendung in der Hochfrequenz- und Ultraschalltech-nik -Stuttgart: Teubner, (1)1951, (2)1953	52	6,432	9	E
Bernal,J.D.	The structure of liquids -San Francisco: Freeman, 1960	9		18	
Berry,L.G. & Mason,B.	Mineralogy -San Francisco: Freeman, 1959	612	13,963	15	U
Berry,L.G. & Thompson,R.M.	X-ray powder data for ore minerals -New York: Geological Society, 1962	281	16,855	5,15,T	
Bersuker,N.B. & Ablov,A.V.	Khimičeskaja svjaz'v kompleksnykh soedinenijakh [Chemical aspects of complex bonding] -Kišinev: Izdat."Štiinca", Akad.Nauk. Moldavskoj SSR, 1962	262		7	
Betekhtin,A.G.	Mineralogija [Mineralogy] -Moskva: NEDRA, 1950	956		15	
[Betekhtin,A.G.=]	Betechtin,A.G. (tr. from Russian, 1950) Lehrbuch der speziellen Mineralogie -Berlin: Verlag Technik, 1957	680		15	U
Betekhtin,A.G.	Kurs mineralogii [Text-book of mineralogy] -Moskva: NEDRA, 1961	539		15	
Bethe,H.A.	Splitting of terms in crystals (tr. of Ann. d. Physik 3, 133, 1929) -New York: Consultants Bureau, 1958	69	11,755	6	
Bhagavantam,S. & Venkataraydu,T.	Theory of groups and its application to physical problems -Waltair (India): Andhra Univ., (1)1948, (2)1951	274	3,79 5,299	6,9,19	
Bhatnagar,S.S. & Mathur,K.N.	Physical principles and applications of magnetochemistry -London: Macmillan, 1935	375		7,9	

Main List - I

Author	Title, publisher, date	No. of pages	Review ref.	Subject ref.	Teaching level
Bijvoet,J.M., Kolkmeijer,N.H. & MacGillavry,C.H.	Röntgenanalyse van kristallen -Amsterdam: Centen, 1938	300	3,322	1	U
Bijvoet,J.M., Kolkmeijer,N.H. & MacGillavry,C.H.	(tr. from Dutch, 1938) Röntgenanalyse von Krystallen -Berlin: Springer, 1940	228		1	U
Bijvoet,J.M., Kolkmeijer,N.H. & MacGillavry,C.H.	(tr. from Dutch, 1938 rev.) X-ray analysis of crystals -London: Butterworths, 1951	304		1	U
Billington,D.S. (ed.)	Radiation damage in solids - see Italian Physical Society, List III			11	
Billington,D.S. & Crawford,J.H.	Radiation damage in solids -Princeton (N.J.): Univ.Press, 1961	462		11	
Birks,J.B. (ed.)	Modern dielectric materials -New York: Academic Press, 1960	253		9	
Birss,R.R.	Symmetry and magnetism -Amsterdam: North Holland, 1964	260		9	
Blackman,M. (ed.)	- see International encyclopedia of physical chemistry and chemical physics				
Bloss, F.D.	An introduction to the methods of optical crystallography -New York: Holt, Rinehart & Winston, 1961	294		14	U
Boas,W.	Physics of metals and alloys -Melbourne: Univ.Press, 1947	193		9,12	
Boer,J.H. de (ed.)	Reactivity of solids - see Conference 1960, List II				
Bogomolov,S.A.	Vyvod pravil'nykh sistem po metody Fedorova [Deduction of crystallographic groups by Fedorov's method] -Leningrad: KUBUCH, Č.1 [Pt.1] 1932 Č.2 [Pt.2] 1934	100 192		2	
Bokij,G.B.	Kristallooptičeskij analiz [Analysis by crystal optics] -Moskva: NAUKA, 1944	156		14	
Bokij,G.B.	Immersionnyj metod [Immersion methods] -Moskva: Izdat.Univ., 1948			14	
Bokij,G.B.	Vvedenie v kristallokhimiju [Introduction to crystal chemistry] -Moskva: Izdat.Univ., 1954	490		7	
Bokij,G.B.	Kristallokhimija [Crystal chemistry] -Moskva: Izdat.Univ., 1961	357		7	
Bokij,G.B. (ed.)	- see Kristalličeskie struktury, List III				

Main List - I

Author	Title, publisher, date	No. of pages	Review ref.	Subject ref.	Tea in lev
Bokij,G.B. & Porai-Koshits,M.L. [See also Porai-Koshits,M.L.]	Praktičeskij kurs rentgenostrukturnogo analiza, Tom I [Practical course of X-ray analysis, Vol.I] -Moskva: Izdat.Univ., 1951	429	6,573	1	
Boldyrev,A.K.	Kristallografija [Crystallography] -Leningrad: KUBUCH, (1)1930, (3)1934	431		1	
Boldyrev,A.K. (& others: see also Dolivo-Dobrovol'skii)	Opredelitel' kristallov, Tom 1, 1-ja polovina: tetragonal'nye kristally [Handbook of crystals, Vol.1, section 1: tetragonal crystals] -Leningrad: NEDRA, 1937	438		T	
Boldyrev,A.K., Mikheev,V.I. & Dubinina,V.N.	Tablici mežploskostnykh rasstojanij dlja železnogo, mednego i molibdenovogo antikatodov [Tables of spacings for Fe, Cu and Mo anticathodes] -Leningrad: Metallurgija, 1950	275		T,5	
Bonštedt,E.M.	Rukovodstvo po izmereniju i vyčisleniju kristallov po metodu Gol'dšmidta [Measurement and calculation of crystals by Goldschmidt's method] -Leningrad: NAUKA, 1934	125		2	
Booth,A.D.	Fourier technique in X-ray organic structure analysis -Cambridge (England): Univ.Press, 1948	106	2,132	3	
Borasky,R. (ed.)	Ultrastructure of protein fibres - see Conference 1961, List II				
Born,M.	Atomtheorie des festen Zustandes -Leipzig: Teubner, 1923			9	U
Born,M. & Huang,K.	Dynamical theory of crystal lattices -Oxford: Univ.Press, 1954	420	9,837	9	G
Bouman,J. (ed.)	Selected topics in X-ray crystallography from the Delft X-ray institute -Amsterdam: North Holland, 1951	375	5,156	1	
Bowden,F.P. & Yoffe,A.D.	Fast reactions in solids -London: Butterworths, 1958	164		16	
Bradley,W.F. (ed.)	- see Clays and clay minerals, List III				
Bragg,W.H.	An introduction to crystal analysis -London: Bell, 1928	168		1	
Bragg,W.H.	(tr. from English, 1928) Vvedenie v analiz kristallov [Introduction to crystal analysis] -Moskva: Gosizdat., 1930	224		1	
Bragg,W.H. & Bragg,W.L.	X-rays and crystal structure -London: Bell, (1)1915, (4)1924	322		1	

Main List - I

Author	Title, publisher, date	No. of pages	Review ref.	Subject ref.	Teaching level
Bragg, W.H. & Bragg, W.L.	(tr. from English, 1924) Rentgenovskie luči i stroenie kristallov [X-rays and crystal structure] -Moskva: Gosizdat., 1929	268		1	
Bragg, W.H., Laue, M. von, & Hermann, C. (eds.)	- see Internationale Tabellen, List III				
Bragg, W.L.	The structure of silicates -Leipzig: Akademische Verlagsg. (1)1930, (2)1932	78		8,15	
Bragg, W.L. (& Schiebold, E.; tr. Brodersen, E.K.; ed. Padurov, N.N., from English, 1931)	Struktura silikatov; s. pril. st. Schiebolda o strukture polevykh špatov [The structure of silicates; with appendix by Schiebold on the structure of felspar] -Moskva: NEDRA, 1934	127		8,15	
Bragg, W.L.	The crystalline state, Vol.1: a general survey -London: Bell, (1)1933, (corr.)1949	352		1	
Bragg, W.L.	(tr. from English, 1933) Kristalličeskoe sostojanie, Tom 1: Obščii obzop [The crystalline state, Vol.1: A general survey] -Moskva: GNTI, 1938	336		1	
Bragg, W.L.	Atomic structure of minerals -Ithaca (N.Y.): Cornell Univ.Press, 1937	192		7,15	U
Bragg, W.L.	The history of X-ray analysis -London: Longmans, 1943	25	20		
Brand, J.C.D. & Speakman, J.C.	Molecular structure: the physical approach -London: Arnold, 1960	300		7,9	
Brandenberger, E.	Angewandte Kristallstrukturlehre -Berlin: Borntraeger, 1938	208		7	UG
Brandenberger, E.	Röntgenographisch-analytische Chemie -Basel: Birkhäuser, 1946	287		7	UG
Brandenberger, E. & Epprecht, W.	Röntgenographische Chemie -Basel: Birkhäuser, (2) 1960	262		7	U
Brasseur, H.	Structure moléculaire des corps solides -Paris: Hermann, 1939	74		3,7	G
Brasseur, H.	Les rayons X et leurs applications -Liège: Desoer, 1945	366		1	U
Braude, E.A. & Nachod, F.C. (eds.)	Determination of organic structures by physical methods -New York: Academic Press, 1955	810	19		
Brauns, R. & Chudoba, K.F.	Allgemeine Mineralogie -Berlin: De Gruyter, (1)1893, (11)1963	152	15		EU

Main List - I

Author	Title, publisher, date	No. of pages	Review ref.	Subject ref.	Tea in lev
Brauns,R. & Chudoba,K.F.	Spezielle Mineralogie -Berlin: De Gruyter, (1)1893, (11)1964	194		15	E
Bravais,M.A.	(tr. Shaler,A.J. from French, 1850) On the systems formed by points regularly distributed on a plane or in space (Journal de L'Ecole Polytechnique, 1850, 19, 1) -American Crystallographic Association, 1949	113	4,80	2,20	
Bridgman memorial volume - see Paul & Warschauer, (eds.)					
Brill,R. (ed.)	- see Fortschritte der Strukturforschung mit Beugungsmethoden (Advances in structure research by diffraction methods), List III				
Brillouin,L.	La structure des corps solides dans la physique moderne -Paris: Hermann, 1937	56		9	C
Brillouin,L.	Wave propagation in periodic structures: electric filters and crystal lattices -New York: McGraw Hill, 1946	259		9	U
Brillouin,L. & Parodi,M.	Propagation des ondes dans les milieux périodiques -Paris: Masson, 1956	347	9,691	9	U
Brindley,G.W. (ed.)- see Brown,G. (ed.)					
Broglie,L.de	Les applications de la mécanique ondulatoire - see Conference, 1953, List II				
Broglie,M.de	Les rayons X -Paris: Blanchard, 1922	164		1	
Brown,G. (ed.)	The X-ray identification and crystal structures of clay minerals -London: Mineralogical Society, (1) (ed. Brindley,G.W.) 1951, (2)1961	544	5,155	15,T	
Brown,N.R. (ed.)	The fibrous proteins - see Conference, 1954, List II				
Bruhns,W. & Ramdohr,P.	Kristallographie -Berlin: De Gruyter, (1)1926, (5)1958	109	14,442	1	E
Buchwald,E.	Einführung in die Kristalloptik -Berlin: De Gruyter, (1)1912, (5)1963	250	5,857 17,1090	14	E
Buckley,H.E.	Crystal growth -London: Chapman & Hall, 1951	571	5,856	13	
Bueren,H.G.van	Imperfections in crystals -Amsterdam: North Holland, (1)1960, (2)1961	676		11	
Buerger,M.J.	Numerical structure factor tables (Special paper 33) -New York: Geological Society of America, 1941, repr.1960	119		T,2	

Main List - I

Author	Title, publisher, date	No. of pages	Review ref.	Subject ref.	Teaching level
Buerger,M.J.	X-ray crystallography -New York: Wiley, (1)1942, repr.1958	531		1	UG
Buerger,M.J.	(tr. Tarasova,V.P. & Shaskolskaya,M.P.; ed. Umanskij,M.M., from English, 1942) Rentgenovskaja kristallografija [X-ray crystallography] -Moskva: MIR, 1948	484		1	
Buerger,M.J.	The photography of the reciprocal lattice (ASXRED monograph No.1) -American Crystallographic Association, 1944	37		5	
Buerger,M.J.	Elementary crystallography -New York: Wiley, 1956	528	10,387	2	U
Buerger,M.J.	Vector space -New York: Wiley, 1959	347	13,280	2	G
Buerger,M.J.	(tr. Belova,E.N. & Simonov,V.I.; ed. Belov,N.V.) Struktura kristallov i vektornoe prostranstvo [Vector space] -Moskva: MIR, 1961	384		2	
Buerger,M.J.	Crystal-structure analysis -New York: Wiley, 1960	668		1	G
Buerger,M.J.	The precession method in X-ray crystallography -New York: Wiley, 1964	288		5	
Bunn,C.W.	Chemical crystallography -Oxford: Univ.Press, (1)1945, (2)1961	509	1,227 15,623	1,7	U
Bunn,C.W.	Crystals (their role in nature and science) -New York: Academic Press, 1964	286		1	E
Burckhardt,J.J.	Die Bewegungsgruppen der Kristallographie -Basel: Birkhäuser, 1947	186	1,46	2,6	G
Burri,C.	Das Polarisationsmikroskop -Basel: Birkhäuser, 1950	308	4,384	14	
Cady,W.G.	Piezoelectricity -New York: McGraw Hill, 1946 -New York: Dover (rev.), 1963	806 822		9	G
Cagnet,M., Françon,M. & Thrierr,J.C.	Atlas optischer Erscheinungen (Atlas de phenomènes d'optique; atlas of optical phenomena) -Berlin: Springer, 1962	91		14	
Callaway,J.	Electron energy bands in solids [reprint from Solid state physics 1958] -New York: Academic Press, 1964	114		9	
Cauchois,Y.	Les spectres de rayons X et la structure électronique de la matière -Paris: Gauthier-Villars, 1948	97		19	
Cauchois,Y. (ed.)	Actions des rayonnements de grande énergie sur les solides - see Conference 1956, List II				

Main List - I

Author	Title, publisher, date	No. of pages	Review ref.	Subject ref.	Te i le
Cauchois,Y. & Hulubei,H.	Longueurs d'onde des émissions X et des discontinuités d'absorptions X -Paris: Hermann, 1947	200	2,131	5,19,T	
Cement and Concrete Association	Chemistry of cement - see Conference 1952,List II				
Chalmers,B.	Principles of solidification -New York: Wiley, 1964	319		16,10	
Chalmers,B. (ed.)	- see Progress in metal physics, List III & Progress in material science, List III				
Champion,F.C.	Electronic properties of diamond -London: Butterworths, 1963	132		8,9	
Chaskolskaia,M.	- see Shaskolskaya,M.				
Chemical Society	Tables of interatomic distances - see Sutton,L.E. (ed.)				
Christie,O.H.J. (ed.)	Feldspars - see Conference 1962, List II				
Chudoba,K.F. (ed.)	- see Hintze, List III				
Cinzerling,E.V.	Iskustvennoe dvoinikovanie kvarca [Artificial twinning of quartz] -Moskva: NAUKA, 1961	158		13	
Clark,G.L.	Applied X-rays -New York: McGraw Hill, (1)1927, (3)1940	674		1	
Clark,G.L.	The encyclopaedia of X-rays and gamma rays -New York: Reinhold, 1963	1149		1	
Commissariat à l'énergie atomique	Diffusion de l'état solide - see Conference 1958, List II				
Correns,C.W.	Einführung in die Mineralogie -Berlin: Springer, 1949	414	3,402	15	U
Cosslett,V.E., Engström,A. & Pattee,H.H. (eds.)	X-ray microscopy and microradiography - see Conference 1956, List II				
Cosslett,V.E. & Nixon,W.C.	X-ray microscopy -Cambridge (England): Univ.Press, 1960	420		19	
Cottrell,A.H.	Theoretical structural metallurgy -London: Arnold, (1)1948, (2)1955	251		12	U
Cottrell,A.H.	(tr. from English) Metallurgie structurale théorique -Paris: Dunod, 1955	330		12	U
Cottrell,A.H.	Dislocations and plastic flow in crystals -Oxford: Univ.Press, 1953	223		11	U
Cottrell,A.H.	The mechanical properties of matter -New York: Wiley, 1964	430		9	
Coulson,C.A.	Valence -Oxford: Univ.Press, (1)1952, (2)1961	404	17,464	7	U

Main List - I

Author	Title, publisher, date	No. of pages	Review ref.	Subject ref.	Teaching level
Cullity,B.D.	Elements of X-ray diffraction -Reading (Mass): Addison-Wesley, 1956	514	10,589	5	U
Cundy,H.M. & Rollett,A.P.	Mathematical models -Oxford: Univ.Press, (1)1951, (2)1961	284		2,21	
Curie,D.	Luminescence cristalline -Paris: Dunod, 1960	206		9	
Curie,D.	(tr. Garlick,G.F.J., from French 1960) Luminescence in crystals -London: Methuen, 1963	332	16,1262	9	
Czechoslovak Academy of Sciences	Semi-conductor physics - see Conference 1960, List II				
Dana,E.S. (& Hurlbut,C.S.)	Minerals and how to study them -New York: Wiley, (1)1895, (3)1949	323	3,78	15	E
Dana,E.S. (& Hurlbut,C.S.)	Dana's manual of mineralogy -New York: Wiley, (1)1912, (17)1959	609	6,111	15	U
Dana,J.W. & Dana,E.S.	The system of mineralogy - see Dana, List III				
Danilov,V.I.	Rassejanie rentgenovskikh lučej v židkostjakh [Diffraction of X-rays by liquids] -Leningrad: Gostekhizdat, 1935	140		18	
Danilov,V.I.	Stroenie i kristallijacija židkosti [Structure and crystallisation of a liquid] -Kiev: Akad.nauk. Ukr.SSR, 1956	568		18	
D'Arcy Thompson	- see Thompson,D.W.				
Das,T.P. & Hahn,E.L.	Nuclear quadrupole resonance spectroscopy -New York: Academic Press, 1958	200		19	
Dauvillier,A.	La technique des rayons X -Paris: P.U.F., 1924	195		5	
Davey,W.P.	A study of crystal structure and its applications -New York: McGraw Hill, 1934	695		1	
Davydov,A.S.	Teorija pogloščenija sveta v molekuljarnikh kristallakh [Theory of molecular excitons] -Kiev: Akad.nauk. Ukr.SSR, 1951	176		9	
Davydov,A.S.	(tr. Kasha,M. & Oppenheimer,M. from Russian) Theory of molecular excitons -New York: McGraw Hill, 1962	174		9	

Main List - I

Author	Title, publisher, date	No. of pages	Review ref.	Subject ref.	Tea in lev
Deer,W.A., Howie,R.A. & Zussman,J.	Rock-forming minerals -London: Longmans			15	
	Vol.1: Ortho-and ring silicates, 1962	333			
	Vol.2: Chain silicates, 1963	379			
	Vol.3: Sheet silicates, 1962	270			
	Vol.4: Framework silicates, 1963	435			
	Vol.5: Non-silicates, 1962	371			
Deicha,G.	Les lacunes des cristaux et leurs inclusions fluides -Paris: Masson, 1955	126	9,326	13	U
Dekeyser,W. & Amelinckx,S.	Les dislocations et la croissance des cristaux -Paris: Masson, 1955	184	9,772	11,13	U
Dekker,A.J.	Solid state physics -New York: Prentice Hall, 1957	540		9	U
Delone,B., Padurov,N. & Aleksandrov,A.	Matematičeskie osnovy strukturnogo analiza kristall i opredelenija osnovnogo parallelepipeda povtorjaemosti pri pomošči rentgenovskikh lučej [Mathematical introduction to crystal structure analysis and the determination of the unit cell by X-rays] -Leningrad: Gostekhizdat., 1934	328		2	
Deryagin,B.V. (ed.; tr. from Russian)	Research in surface forces -New York: Consultants Bureau, 1963	190	17,75	10	
Dettmar,H.K. & Kirchner,H.	Tabellen zur Auswertung der Röntgendiagramme von Pulvern (Debye-Scherrer Diagramme) -Weinheim: Verlag Chemie, 1956	94	9,1040	T,5	
D'Eye,R.W.M. & Wait,E.	X-ray powder photography -London: Butterworths, 1960	222	15,96	5	
Dienes,G.J. & Vineyard,G.H.	Radiation effects in solids -New York: Interscience, 1957	234		11	
Dienes,G.J. & Vineyard,G.H.	(tr. Breger,A.Kh. from English, 1957) Radiacionnye ėffekty v tverdnkh telakh [Radiation effects in solids] -Moskva: MIR, 1960	243		11	
Dolivo-Dobrovol'skii,V.V.	Kurs kristallografii [A crystallography course] -Leningrad: NEDRA, 1937	347		1	
Dolivo-Dobrovol'skii,V.V. (and others)(see also Boldyrev)	Oppredelitel' kristallov, Tom.1, 2-ja polovina: Geksagonal'nye i trigonal'nye kristally [Handbook of crystals, Vol.1, section 2: Hexagonal and trigonal crystals] -Leningrad: NEDRA, 1939	863		T	

Main List - I

Author	Title, publisher, date	No. of pages	Review ref.	Subject ref.	Teaching level
Donnay,J.D.H.	Spherical trigonometry after the Cesaro method -New York: Interscience, 1945	83		2	U
Donnay,J.D.H., Nowacki,W. & Donnay,C. (eds.)	Crystal data: Part I (Nowacki) Systematic tables Part II (Donnay and Donnay) Determinative tables -American Crystallographic Association, 1954	719	7,607	T	
Donnay,J.D.H. & Donnay,G. (eds.)(with Cox,E.G., Kennard,O. & King,M.V.)	Crystal data: Part II: Determinative tables [A.C.A. Special monograph no.5] -American Crystallographic Association, (2)1963	1302	17,73	T	
Doremus,R.H., Roberts,B.W. & Turnbull,D. (eds.)	Growth and imperfections in crystals - see Conference, 1958, List II				
Dubinin,M.M. (ed.)	Metody issledovanija strukturn v'sokodispersnykh i poristykh tel [Methods of investigation of the structure of highly dispersed and porous materials] -Moskva: NAUKA, 1953, 1958			5,18	
Dubinin,M.M. & Witzmann,H. (eds.)	(tr. Höhne,E. from Russian, 1953) Methoden der Strukturuntersuchung an hochdispersen und porösen Stoffe -Berlin: Akademie Verlag, 1961	248		5,18	
Eitel,W.	Physikalische Chemie der Silikate -Leipzig: Voss, (1)1929, (2)1941	826		8,15	G
Eitel,W.	The physical chemistry of the silicates -Chicago: Univ.Press, 1954	1592		8,15	
Eitel,W.	Thermochemical methods in silicate investigation -New Jersey: Rutgers Univ.Press, 1952	132		8,15	
Eitel,W.	Structural conversions in crystalline systems and their importance for geological problems (special paper 66) -Geological Society of America, 1958	178		16	
Eitel,W.	Silicate science. Vol.1: Silicate structures -New York: Academic Press, 1964	666		8	
Elcock,E.W.	Order-disorder phenomena -London: Methuen, 1956	166	10,389	11	UG
Engström,A.	X-ray microanalysis in biology and medicine -Amsterdam: Elsevier, 1962	92		17	
Engström,A., Cosslett,V.E. & Pattee,H.H. (eds.)	X-ray microscopy and X-ray microanalysis - see Conference 1959, List II				

Main List - I

Author	Title, publisher, date	No. of pages	Review ref.	Subject ref.	Tea in lev
Engström,A. & Finean,J.B.	Biological ultrastructure -New York: Academic Press, 1958	326	12,352	17	
Enz,W.	Strukturformel und Valenz -St. Gallen: Fehrsche, 1960	50		7	U
Escher,M.C.	Grafiek en tekeningen -Zwolle: van de Erven, 1960	63		21	
Escher,M.C.	(tr. from Dutch, 1960) The graphic work of M.C.Escher -London: Oldbourne, 1961	63		21	
Eskola,P.	Kristalle und Gesteine -Berlin: Springer, 1946	397		15	U
Evans,R.C.	An introduction to crystal chemistry -Cambridge (England): Univ.Press, (1)1939, (2)1964	410	17,1618	7	U
Evans,R.C.	(tr. Makarov,E.S. from English, 1939) Vvedenie v kristallokhimiju [Introduction to crystal chemistry] -Moskva: Khimija, 1948	368		7	
Evans,R.C.	(tr. from English, 1939) Chimie et structure cristalline -Paris: Dunod, 1954	331		7	
Evans,R.C.	(tr. from English, 1939) Einführung in die Kristallchemie -Leipzig: Barth, 1954	307		7	U
Ewald,P.P.	Kristalle und Röntgenstrahlen -Berlin: Springer, 1923	327		4	U
Ewald,P.P. (ed.)	Flüssige Kristalle (reprinted from Z. Krist. vol. 79) -Leipzig: Akademische Verlagsg, 1931	348		18	G
Ewald,P.P. (ed.)	Fifty years of X-ray diffraction -Utrecht: Oosthoek, 1962	771		20	
Ewald,P.P. & Hermann,C. (eds.)	- see Strukturbericht I, List III				
Faddeev,D.K.	Tablicy osnovnykh unitarnykh predstavlenij fedorovskikh grupp [Tables of fundamental unitary representations of the Fedorov groups] -Moskva: NAUKA, 1961	173		T,2,6	
Faraday Society	Crystal growth - see Conference 1949, List II				
Fast,J.D. (ed.)	La diffusion dans les métaux - see Conference 1956, List II				
Fedorov,E.S.	(ed. Anšeles,O.M. & others) Načala učenija o figurakh [Principles of morphology] -Moskva: NAUKA, (1)1886, rep.1953	410		2	
Fedorov,E.S.	Kratkoe rukovodstvo po kristallografii, č.1 [Concise handbook of crystallography,Pt.1] -St.Petersburg: 1891	98		1	

Main List - I

Author	Title, publisher, date	No. of pages	Review ref.	Subject ref.	Teaching level
Fedorov,E.S.	Kurs kristallografii [Crystallography course] -St.Petersburg: (2)1897, (3)1901	438		1	
Fedorov,E.S.	Das Kristallreich: Tabellen zur kristallochemische Analyse -Petrograd: NAUKA, 1920 Text Atlas	 1050 213		 T,2	
Fedorov,E.S.	(ed. Shubnikov,A.V. & Shafranovskii,I.I.) Simmetrija i struktura kristallov [Crystal symmetry and structure] -Moskva: NAUKA, 1949	630		2	
Fedorov,E.S.	[Universal stage: collected articles] - see Zavarickij, (ed.)				
Fedorov,F.I.	Optika anisotropnykh sred [Optics of anisotropic media] -Minsk: Akad.nauk. BSSR, 1958	380		14	
Fermi,E.	Moleküle und Kristalle -Leipzig: Barth, 1938	234		9,7	
Fermi,E.	(tr.; ed. Beresteckij,V.B.) Molekuly i kristally [Molecules and crystals] -Moskva: MIR, 1947	266		9,7	
Fersman,A.E.	Očerki po istorii kamnja [Notes on the history of stones] -Moskva: NAUKA Tom 1, 1954 Tom 2, 1961	 372 371		15,21	
Fersman,A.E.	Vospominanija o kamne [Recollections about stones] -Moskva: NAUKA, 1960	167		15,21	
Fersman,A.E.	(ed. Beljankin,D.S. & Shafranovskii,I.I.) Kristallografija almaza [The crystallography of diamond] -Moskva: NAUKA, 1955	566		15,8	
Fersman,A.von & Goldschmidt,V.	Der Diamant -Heidelberg: Carl Winter, 1911	566		15,8	
Fischer,E.	Einführung in die geometrische Kristallographie -Berlin: Akademie Verlag, 1955	162		2	EU
Fisher,J.C., Johnston,W.G., Thompson,R. & Vreeland,T. (eds.)	Dislocations and mechanical properties of crystals - see Conference 1956, List II				
Fletcher,L.	The optical indicatrix and the transmission of light through crystals -London: Frowde, 1892	112		14,20	
Flint,E.E.	Praktičeskoe rukovodstvo po geometričeskoj kristallografii [A practical handbook of geometrical crystallography] -Moskva: NEDRA, (1)1937, (3)1956	208		2	

Main List - I

Author	Title, publisher, date	No. of pages	Review ref.	Subject ref.	Tea in lev
Flint,E.E.	Načala kristallografii [Principles of crystallography] -Moskva: NEDRA (1)1952, (2)1961	224		1	
Flint,E.E. & Anvaer,A.D.	Stereoal'bom po kristallografii [Stereo-album of crystallography] -Moskva: Prosveščenie Pt.1, 1939 Pt.2, 1940	 64 100		 1	
Flügge,S. (ed.)	- see Handbuch der Physik, List III				
Flügge,S. & Trendelenburg,F. (eds.)	- see Ergebnisse der exakten Naturwissenschaften, List III				
Fox,D. & others	Physics and chemistry of the organic solid state (Vol.1) -New York: Interscience, 1963	823		7,9	
Francombe,M.H. & Sato,H. (eds.)	Single crystal films - see Conference 1963, List II				
Frank-Kameneckij,V.A., Kondrat'eva,V.V. & Kamencev,I.E.	Rukovodstvo dlja laboratornykh zanjatij po rentgenovskomu issledovaniju mineralov [Laboratory handbook for X-ray investigation of minerals] -Leningrad: Izdat.Univ., 1959	199		5,15	
Fréchette,V.D. (ed.)	Non-crystalline solids - see Conference 1958, List II				
Friedel,G.	Étude sur les groupements cristallins -St.Étienne: Société de l'imprimerie Théolier, J.Thomas et Cie., 1904	481		2	
Friedel,G.	Leçons de cristallographie -Paris: Hermann, (1)1911 -Paris: Berger-Levrault, (2)1926 -Paris: Blanchard, repr. 1964	 310 601		 1	 U
Friedel,J.	Les dislocations -Paris: Gauthier-Villars, 1956	314	10,488	11	
Friedel,J.	(tr. Vassamillet,L.F., from French, 1956 rev.) Dislocations -Oxford: Pergamon, 1964	491		11	
Friedel,J. & Guinier,A. (eds.)	Metallic solid solutions - see Conference 1962, List II				
Frisch,O.R., Paneth,F.A., Laves,F. & Rosbaud,P. (eds.)	Beiträge zur Physik und Chemie des 20 Jahrhunderts: Lise Meitner, Otto Hahn, Max von Laue zum 80 Geburtstag -Braunschweig: Vieweg, 1959	285		20	
Furnas,T.C.	Single crystal orienter instruction manual -Milwaukee (Wis.): General Electric Co., 1957	174	13,170	5	

Main List - I

Author	Title, publisher, date	No. of pages	Review ref.	Subject ref.	Teaching level
Gadolin,A.V.	(ed. Anšeles,O.M. & others) Vyvod vsekh kristallografičeskikh sistem i ikh podrazdelenii iz odnogo obščego načala [Deduction of all crystal systems and their subdivisions from one general principle] -Moskva: NAUKA, 1954	157		20,2	
Garner,W.E.	Chemistry of the solid state -London: Butterworths, 1955	417		7,9	
Garratt,A. (ed.)	Penguin science survey, 1961 -Harmondsworth (England): Penguin Books, 1961	239	15,425	1	EU
Garrido,J.	Problemas de cristalografia (morfologica y estructural) -Madrid: Ediciones Hispano-Argentinas, 1949	149	3,323	2	U
Garrido,J.	Leçons sur la structure atomique des cristaux -Porto: Instituta para a alta cultura, 1951	148		1	
Garrido,J. & Orland,J.	Los rayos X y la estructura fina de los cristales: fundamentos teoricos y metodos practicos -Madrid: Editorial Dossat, 1946	260	1,48	1	
Gay,H.	Cours de cristallographie -Paris: Gauthier-Villars			1	U
	Livre I: Cristallographie géométrique, 1958	253	14,331		
	Livre II: Cristallographie physico-chimique, 1959	232	14,331		
	Livre III: Radiocristallographie théorique, 1961	278			
Geil,P.H.	Polymer single crystals -New York: Interscience, 1963	560		17	
Gercriken,S.D., Dekhtjar',I.Ja., Krivoglaz,M.A. & others	Fizičeskie osnovy pročnosti i plastičnosti metallov [Physical principles of strength and plasticity of metals] -Moskva: Metallurgija, 1963	322		12,9	
Gilman,J.J.	The art and science of growing crystals -New York: Wiley, 1963	493		13	
Glass (All-union Conference on) - see Glassy state, List III					
Glasser,O.	Wilhelm Conrad Röntgen und die Geschichte der Röntgenstrahlen -Berlin: Springer, (1)1931, (2)1959	381		20	
Glocker,P.R.	Materialprüfung mit Röntgenstrahlung -Berlin: Springer, (1)1936, (4)1958	530	4,78 11,702	1	UG
Glocker,P.R.	(tr. from German) Rentgenovskie luči i ispytanie materialov [Materialprüfung mit Röntgenstrahlung] -Moskva: Gostekhizdat., 1932	396		1	

Main List - I

Author	Title, publisher, date	No. of pages	Review ref.	Subject ref.	Te i le
Goldschmidt,V.	Kristallographische Winkeltabellen -Berlin: Springer, 1897	432		T	
Goldschmidt,V.	Atlas der Krystallformen -Heidelberg: Carl Winter, 1913-26 [9 volumes of figures, & 9 volumes of tables]			T,13	
Goldschmidt,V.M.	Geochemische Verteilungsgesetze der Elemente (Skrifter utgit av Det Norske Videnskaps-Akademi) -Oslo: Jacob Dybwad			7,15	
	I. 1923	17			
	II. Beziehungen zwischen den geochemischen Verteilungsgesetzen und dem Bau der Atome, 1924	38			
	III. Röntgenspektrographische Untersuchungen über die Verteilung der seltenen Erdmetalle in Mineralien, [with L.Thomassen], 1924	58			
	IV. Zur Krystallstruktur der Oxyde der seltenen Erdmetalle [with F.Ulrich & T.Barth], 1925	24			
	V. Isomorphie und Polymorphie der Sesquioxyde. Die Lanthaniden-Kontraktion und ihre Konsequenzen [with T.Barth & G.Lunde], 1925	59			
	VI. Über die Krystallstrukturen vom Rutiltypus, mit Bemerkungen zur Geochemie zweiwertiger und vierwertiger Elemente [with T.Barth, D.Holmsen, G.Lunde & W.Zachariasen], 1926	21			
	VII. Die Gesetze der Krystallochemie [with T.Barth, G.Lunde, & W.Zachariasen], 1926	117			
	VIII. Untersuchungen über Bau und Eigenschaften von Krystallen, 1927	156			
	IX. Die Mengenverhältnisse der Elemente und der Atom-Arten, 1938	148			
Goldschmidt,V.M.	(tr. Borneman-Starynkevič,I.D. & others, from German) Raboty po geokhimikii i kristallokhimii 1911-30 [Work on geochemistry and crystal chemistry 1911-30] -Leningrad: Khimija, 1933	275		7,15	
Goldschmidt,V.M.	(tr. Borneman-Starynkevič,I.D.) Kristallokhimija (Crystal chemistry) -Leningrad: Khimija, 1937	63		7,15	
Goldschmidt,V.M.	(ed. Muir,A.) Geochemistry -Oxford: Univ.Press, 1954	742		7,15	
Gomer,R. & Smith,C.S. (eds.)	Structure and properties of solid surfaces - <u>see</u> Conference 1952, List II				
Goodenough,J.B.	Magnetism and the chemical bond -New York: Interscience, 1963	393		7,9	

Main List - I

Author	Title, publisher, date	No. of pages	Review ref.	Subject ref.	Teaching level
Gorelik,S.S., Rastorguev,L.N. & Skakov,Ju.A.	Rentgenografičeskij i ėlektronografičeskij analiz metallov: priloženie otdel'nyij tom atlas tablič i rentgenogramm [Analysis of metals by X-ray diffraction and electron diffraction: with a separate volume of tables and X-ray diagrams -Moskva: Metallurgija, 1963 Text Atlas, tables, diagrams	256 92		5,12,T	
Gorter,G.J. (ed.)	- see Progress in low temperature physics, List III				
Gray,G.W.	Molecular structure and the properties of liquid crystals -New York: Academic Press, 1961	350	16,235	18	
Green,H.S. & Hurst,C.A.	Order-disorder phenomena -New York: Interscience, 1964	363		11,16	
Griffith,J.S.	The theory of transition-metal ions -Cambridge (England): Univ.Press, 1961	466		7	
Grigor'ev,D.P.	Ontogenija mineralov [Ontogenesis of minerals] L'vov: Izdat.Univ., 1961	284		15	
Grigor'ev,D.P.	Osnovy konstitucii mineralov [Principles of mineral constitution] -Moskva: NEDRA, 1962	63		15	
Grim,R.	Clay mineralogy -New York: McGraw Hill, 1953	384		15	
Grim,R.	Applied clay mineralogy -New York: McGraw Hill, 1962	422		15	
Grofcsik,J. & Tamas,F.	Mullite, its structure, formation and significance -Budapest: Akadémiai Kiadó (Publishing House of the Hungarian Academy of Sciences), 1961	164		8,15	
Groth,P.	Physikalische Kristallographie -Leipzig: Engelmann, 1905	820		9	
Groth,P.	Chemische Kristallographie, Bd. 1-5 -Leipzig: Engelmann, 1906-1919			T,7	
Gschneider,K.	Rare earth alloys -New York: Van Nostrand, 1961	500		12,8,T	
Guinier,A.	Théorie et technique de la radiocristallographie -Paris: Dunod, (1)1945, (3)1964	740	10,368	1,5	UG
Guinier,A.	(tr. Tippell,T.L.; ed. Lonsdale,K., from French, 1945, rev.) X-ray crystallographic technology -London: Hilger & Watts, 1952	330	6,751	1,5	UG

Author	Title, publisher, date	No. of pages	Review ref.	Subject ref.	Tea in lev
Guinier,A.	(tr. Belova,E.N., Kvitka,S.S. & Tarasova,B.P.; ed. Belov,N.V.) Rentgenografija kristallov [Théorie et technique de la radiocristallographie] -Moskva: NAUKA, 1961	604		1,5	
Guinier,A.	(tr. Tjapkin,Ju.D.; ed. Bagarjackij,Ju.A.) Neodnorodnye metalličeskie tverdye rastvory [Heterogeneities in metallic solid solutions] -Moskva: MIR, 1962	158		11,12	
Guinier,A.	(tr. Lorrain,P. & Lorrain,D.S., from part of French, 1956, rev.) X-ray diffraction in crystals, imperfect crystals, and amorphous bodies -San Francisco: Freeman, 1963	378	17,1617	4,5	[illegible]
Guinier,A. & Dexter,D.L.	X-ray studies of materials -New York: Interscience, 1963	156		5,1	[illegible]
Guinier,A. & Fournet,G.	(tr. Walker,C.B.) Small-angle scattering of X-rays -New York: Wiley, 1955	268	9,839	5,11	[illegible]
Guttman,L.	Order-disorder phenomena in metals - see Muto, Takagi & Guttman				
Hagan,M.	Clathrate inclusion compounds -New York: Reinhold, 1962	189		8	
Hall,E.O.	Twinning and diffusionless transformations in metals -London: Butterworths, 1954	181	7,524	12,16	
Halla,F.	Kristallchemie und Kristallphysik metallischer Werkstoffe -Leipzig: Barth, (3)1957	737	5,300	12	[illegible]
Harrison,W.A. & Webb,M.B.	The Fermi surface - see Conference 1960, List II				
Hartree,D.R.	The calculation of atomic structures -London: Chapman & Hall, 1957	181	11,376	6	
Hartshorne,N.H. & Stuart,A.	Crystals and the polarizing microscope -London: Arnold, (1)1934, (3)1960	556	13,853	14	[illegible]
Hartshorne,N.H. & Stuart,A.	Practical optical crystallography -London: Arnold, 1964	326		14	
Hassell,O.	Kristallchemie -Dresden: Theodor Steinkopff, 1934	114		7	
Hassell,O.	(tr. Evans,R.C.) Crystal chemistry -London: Heinemann, 1935	94		7	

Main List - I

Author	Title, publisher, date	No. of pages	Review ref.	Subject ref.	Teaching level
Hassell,O.	(tr. Belov,N.V.; ed. Shubnikov,A.V.) Kristallokhimija [Kristallchemie] -Leningrad: Khimija, 1936	200		7	
Hauffe,K.	Reaktionen in und an festen Stoffen -Berlin: Springer, 1955	696		16	G
Hauptman,H. & Karle,J.	Solution of the phase problem (A.C.A. Monograph No.3) -American Crystallographic Association, 1953	87	8,365	3	
Hauy,R.J. [=Gajui,R.Ž.]	(tr.; ed. Shubnikov,A.V. & Bokij,G.V., from French of 1784) Struktura kristallov [Essai d'une théorie sur la structure des crystaux appliquées à plusieurs genres de substances crystallisées] -Leningrad: NAUKA, 1962	176		20	
Hedvall,J.A.	Einführung in die Festkörperchemie -Braunschweig: Vieweg, 1952	304		7,16	
Hein,F.	Chemische Koordinationslehre -Den Haag: Nijhoff, 1957	683		7	
Heller,L. & Taylor,H.F.W.	Crystallographic data for the calcium silicates -London: H.M.S.O., 1956	79		T,8,15	
Henry,N.F.M., Lipson,H. & Wooster,W.A.	The interpretation of X-ray diffraction photographs -London: Macmillan, (1)1951, (2)1960	282	5,859 15,95	5	U
Herman,F. & Skillman,S.	Atomic structure tables -London: Prentice Hall, 1963	448		6,T	
Hermann,C. (ed.)	- see Internationale Tabellen, List III				
Hermann,C., Lohrmann,O. & Philipp,H.	- see Strukturbericht, List III				
Herzog,W.	Oszillatoren mit Schwingkristallen -Berlin: Springer, 1958	317		9	UG
Hevesy,G.von	Chemical analysis by means of X-rays and its applications Ithaca (N.Y.): Cornell Univ.Press, 1932			7	
Hey,M.H.	An index of mineral species and variations arranged chemically -London: British Museum (Natural History), (1)1955, (2)1962	728	4,479	15,T	
Hey,M.H.	An index of mineral species and variations arranged chemically: Appendix -London: British Museum (Natural History), 1963	135		15,T	
Hiller,J.E.	Grundriss der Kristallchemie -Berlin: De Gruyter, 1952	307	6,304	7	U

Author	Title, publisher, date	No. of pages	Review ref.	Subject ref.	Tea in lev
Hilton,H.	Mathematical crystallography, and the theory of groups of movements -Oxford: Univ.Press, 1903 -New York: Dover, 1963	262		2,6	
Himmel,L. (ed.)	Recovery and recrystallisation of metals - see Conference 1962, List II				
Hintze,C.	Handbuch der Mineralogie - see List III				
Hippel,A.E.von (with others)	Molecular science and molecular engineering -New York: Wiley, 1959	446		9	[illegible]
Hirst,H.	X-rays in research and industry -London: Chapman & Hall, (1)1943, (2)1946	124		1	[illegible]
Hocart,R. & Kern,R.	Problèmes et calculs de chimie générale et de cristallochimie -Paris: Gauthier-Villars, 1959	212		7	
Hodgman,C.D. (ed.)	Handbook of chemistry and physics -Cleveland (Ohio): Chemical Rubber Publishing Company, 1962	3520		T	
Hodgman,C.D. (ed.)	Handbook of mathematical tables -Cleveland (Ohio): Chemical Rubber Publishing Company, 1962	590		T	
Holden,A. & Singer,P.	Crystals and crystal growing -New York: Doubleday, 1960	320	14,332	1,13	[illegible]
Holden,A. & Singer,P.	(tr. from English) Die Welt der Kristalle -München: Desch, 1960	315		1,13	[illegible]
Holser,W.T. (ed.)	- see Shubnikov,A.V. & Belov,N.V.				
Honigmann,B.	Gleichgewichts und Wachstumsformen von Kristallen -Darmstadt: Dietrich Steinkopff, 1958	161	12,614	13	[illegible]
Hosemann,R. & Bagchi,S.N.	Direct analysis of diffraction by matter -Amsterdam: North Holland, 1962	734	17,463	4,11	
Houwink,R. (& Burgers,W.G.)	Elasticity, plasticity, and structure of matter -Cambridge (England): Univ.Press, (1)1954 -New York: Dover, repr. 1959	368		9,11	
Houwink,A.L. & Spit,B.J. (eds.)	Electron microscopy - see Conference 1960, List II				
Hückel,W.	Anorganische Strukturchemie -Stuttgart: Enke, 1948	1033		7	
Hückel,W.	Structural chemistry of inorganic compounds -Amsterdam: Elsevier, Vol.I, 1950, Vol.II, 1952	483 656		7	

Main List - I

Author	Title, publisher, date	No. of pages	Review ref.	Subject ref.	Teaching level
Huerta,F.	Los metodos del cristal giratorio -Madrid: C.S.I.C., 1952	108	6,574	5	
Huerta,F.	Teoria de los metodos roentgenograficos del cristal giratorio -Madrid: C.S.I.C., 1955	136	10,540	5	
Hughes,D.J.	Neutron optics -New York: Interscience, 1954	136		5	
Hume-Rothery,W.	Atomic theory for students of metallurgy -London: Institute of Metals, (1)1946, (3)1960	422		7,12	U
Hume-Rothery,W.	Elements of structural metallurgy -London: Institute of Metals, 1961	174		12	U
Hume-Rothery,W., Christian,J.W. & Pearson,W.B.	Metallurgical equilibrium diagrams -London: Institute of Physics, 1952	311		12	
Hume-Rothery,W. & Raynor,G.V.	The structure of metals and alloys -London: Institute of Metals, (1)1936, (4)1962	364		12	U
Huntington,H.B.	The elastic constants of crystals (repr. from Solid state physics 1958) -New York: Academic Press, 1964	140		9	
Hurlbut,C.S. (ed.) - see Dana					
Hutchison,T.S. & Baird,D.C.	The physics of engineering solids -New York: Wiley, 1963	368		9	
International Union of Crystallography - see IUCr.					
International Union of Pure and Applied Chemistry - see IUPAC					
[Ioffe,A.F.=] Joffe,A.F.	The physics of crystals -New York: McGraw Hill, 1928	198		9	
Ioffe,A.F.	Fizika kristallov [Crystal physics] -Moskva: Gosizdat, 1929	192		9	
Ioffe,A.F.	Fizika poluprovodnikov [Physics of semiconductors] -Moskva: NAUKA, 1957	491		9	
[Ioffe,A.F.=] Joffe,A.F.	(tr. from Russian, 1957) Physik der Halbleiter -Berlin: Akademie Verlag, (1)1958, (2)1960	436		9	
Ioffe,A.F.	(tr. Goldsmid,H.J. from Russian, 1957) Physics of semiconductors -London: Infosearch, 1960	436		9	
Italian Physical Society - see List III					
Ito,T.	X-ray studies in polymorphism -Tokyo: Maruzen, 1950	236	5,297	15	

Main List - I

Author	Title, publisher, date	No. of pages	Review ref.	Subject ref.	Tea in lev
IUCr	Index of manufacturers - see Rose,A.J. (ed.)				
IUCr	International tables - see List III				
IUCr	Structure reports - see List III				
IUCr	World directory of crystallographers - see Smits,D.W. (ed.)				
IUPAC	Manual of physico-chemical symbols and terminology -London: Butterworths, 1959	27		T,7	
IUPAC	Rules for notation for organic compounds -London: Longmans, 1961	107		T,7	
Jaeger,F.M.	Lectures on the principle of symmetry and its application in all natural sciences -Amsterdam: Elsevier, (1)1917, (2)1920	348		2,21	
James,R.W.	X-ray crystallography -London: Methuen, (1)1930, (5)1953, corr. 1961	101		1	E
James,R.W.	(tr. from English, 1930) Vvedenie v rentgenovskij analiz [X-ray crystallography] -Moskva: Gostekhizdat., 1932	91		1	
James,R.W.	The optical principles of the diffraction of X-rays (The crystalline state, Vol.2) -London: Bell, (1)1948, (2)1963	640	3,332	4	U
James,R.W.	(tr.; ed. Iveronova,V.I.; from English, 1948) Optičeskie principy difrakcii rentgenovskikh lučej [The optical principles of the diffraction of X-rays] -Moskva: MIR, 1950	572		4	
Jaswon,M.A.	The theory of cohesion -London: Pergamon, 1954	245	9,204	9	
Jaswon,M.A.	Studies in crystal physics -London: Butterworths, 1959	42	13,364	9	
Jaynes,E.T.	Ferroelectricity -Princeton: Univ.Press, 1953	136		9	
Jennison,R.C.	Fourier transforms and convolutions for the experimentalist -Oxford: Pergamon, 1961	120		6	
Joffe,A.F.	- see Ioffe,A.F.				
Johannsen,A.	Manual of petrographic methods -New York: McGraw Hill, (1)1914, (2)1918	648		14	
Jona,F. & Shirane,G.	Ferroelectric crystals -Oxford: Pergamon, 1962	402		9	
Jones,G.O.	Glass -London: Methuen, 1956	119		18	

Main List - I

Author	Title, publisher, date	No. of pages	Review ref.	Subject ref.	Teaching level
Jones,H.	Theory of Brillouin zones and electronic states in crystals -Amsterdam: North Holland, 1960	268		9	
Jong,W.F.de	Compendium der Kristalkunde -Utrecht: Oosthoek, 1951	260	5,858	1	U
Jong,W.F.de	(tr. from Dutch) Kompendium der Kristallkunde -Berlin: Springer, 1959	258	12,613	1	U
Jong,W.F.de & Bouman,J.	General crystallography -San Francisco: Freeman, 1959	281	13,854	1	U
Justi,E.	Leitfähigkeit und Leitungmechanismus fester Stoffe -Göttingen: Vanderhoeck & Rupprecht, 1948	348		9	G
Känzig,W.	Ferroelectrics and antiferroelectrics [repr. from Solid state physics 1957] -New York: Academic Press, 1964	198		9	
Kapustin,A.P.	Vlijanie ul'trazvuka na kinetiku kristalližazii [Effect of ultrasonic radiation on the kinetics of crystallisation] -Moskva: NAUKA, 1962	107		16,13	
Kapustin,A.P.	(tr. from Russian, 1962) Effects of ultrasonic energy on the kinetics of crystallization -New York: Consultants Bureau, 1963	65		16,13	
Ketelaar,J.A.A.	De chemische bindung -Amsterdam: Elsevier, 1952	373		7	UG
Ketelaar,J.A.A.	(tr. from Dutch) Chemical constitution -Amsterdam: Elsevier, (1)1953, (2)1958	448		7	
Ketsmets,D.I.	Kristallografija i mineralogija [Crystallography and mineralogy] -Kharkov: GNTI, 1957	152		15	
Khejker,D.M. & Zevin,L.C. (ed. Ždanov,G.S.)	Rentgenovskaja difraktometrija [X-ray diffractometry] -Moskva: NAUKA, 1963	380		5	
Kitaigorodskii,A.I. (ed.)	Spravočnik po rentgenostrukturnomu analizu [Tables for X-ray structure analysis) -Moskva: Gostekhizdat., 1940	316		T	
Kitaigorodskii,A.I.	Rentgenostrukturnyi analiz [X-ray analysis] -Moskva: Gostekhizdat., 1950	650	6,573	1	
Kitaigorodskii,A.I.	Rentgenostrukturnyi analiz melkokristalličeskikh i amorfnykh tel [X-ray structure analysis of finely-crystallised and amorphous bodies] -Moskva: Gostekhizdat, 1952	583		5	

Author	Title, publisher, date	No. of pages	Review ref.	Subject ref.	Te in le
Kitaigorodskii,A.I.	Porjadok i besporjadok v mire atomov [Order and disorder in the world of atoms] -Moskva: NAUKA, (1)1954, (3)1959	150		1	
Kitaigorodskii,A.I.	Organičeskaja kristallokhimija [Organic crystal chemistry] -Moskva: NAUKA, 1955	559		7	
Kitaigorodskii,A.I.	(tr. from Russian, 1955) Organic chemical crystallography -New York: Consultants Bureau, 1961	541	15,622	7	
Kitaigorodskii,A.I.	Teorija strukturnogo analiza [Theory of structure analysis] -Moskva: NAUKA, 1957	283	12,482	3	
Kitaigorodskii,A.I.	(tr. Harker,D. & Harker,K. from Russian, 1957) The theory of crystal structure analysis -New York: Consultants Bureau,1961	275	15,516	3	
Klassen-Neklyudova,M.V. (ed.)	Nektorye voprosy fiziki plastičnosti kristallov [Some aspects of the plasticity of crystals] -Moskva: NAUKA, 1960	210		9,11	
Klassen-Neklyudova,M.V. (ed.; tr. from Russian, 1960)	Plasticity of crystals -New York: Consultants Bureau, 1962	196	17,622	9,11	
Klassen-Neklyudova,M.V.	Mekhaničeskoe dvojnikovanie kristallov [Mechanical twinning of crystals] -Moskva: NAUKA, 1960	261		13,9	
Klassen-Neklyudova,M.V. (tr. from Russian, 1960)	Mechanical twinning of crystals -New York: Consultants Bureau, 1963	227		13,9	
Kleber,W.	Angewandte Gitterphysik -Berlin: de Gruyter, (1)1941, (3)1960	291	15,172	9	
Kleber,W.	Einführung in die Kristallographie -Berlin: VEB Technik, (4)1961, (7)1964	418		1	
Kleber,W.	Kristallchemie -Leipzig: Teubner, 1963	128	17,793	7	
Klockmann	- see Ramdohr				
Klug,H.P. & Alexander,L.E.	X-ray diffraction procedures for polycrystalline and amorphous materials -New York: Wiley, 1954	715	8,366	5	
Knaggs,I.E., Karlik,B. & Elam,C.F.	Tables of cubic crystal structure of elements and compounds -London: Hilger, 1932	92		T	
Kochendörfer,A.	Plastische Eigenschaften von Kristallen und metallischen Werkstoffe -Berlin: Springer, 1941	312		9,12	

Main List - I

Author	Title, publisher, date	No. of pages	Review ref.	Subject ref.	Teaching level
Kochendörfer,A. & Seeger,A.	Plastizität und Festigkeit der Kristalle und metallischen Werkstoffe -Berlin: Springer, (2)1962			9,12	
König,R.	Anorganische Pigmente und Röntgenstrahlen -Stuttgart: Enke, 1956	132	10,676	T,5	
Koerber,G.G.	Properties of solids -New York: Prentice Hall, 1962	286		9	U
Kohlhaas,R. & Otto,H.	Röntgen-Strukturanalyse von Kristallen -Berlin: Akademie Verlag, 1955	212	9,203	1	U
Kordes,E. (& Recker,K.)	Optische Daten zur Bestimmung anorganischer Substanzen mit dem Polarisationsmikroskop -Weinheim: Verlag Chemie, 1960	192		T,14	
Kordes,E.	Farbige Bilder zur Kristalloptik (entnommen aus "Optische Daten") -Weinheim: Verlag Chemie, 1960	6		14	
Korenman,I.M.	Mikrokristalloskopija [Microscope identification of crystals] -Moskva: Khimija, (1)1947, (2)1955	432		14	
Koster,G.F.	Space groups and their representations [repr. from Solid state physics 1957] -New York: Academic Press, 1964	84		2,6	
Koster,G.F., Dimmock,J.O., Wheeler,R.G. & Statz,H.	Properties of the 32 point groups -Cambridge (Mass.): M.I.T.Press, 1963	104		2	
Kovalev,O.V.	Neprivodimye predstavlenija prostranstvennykh grupp [Irreducible representations of space groups] -Kiev: Akad.nauk.Ukr.SSR, 1961	154		6	
Kožina,M.I., Stroganov,E.V. & Tolkačev,S.S.	Rukovodstvo k laboratornym rabotam po strukturnoj kristallografii [Laboratory handbook of structural crystallography] -Leningrad: Izdat.Univ. Č.1 (Pt.I), 1957 Č.2 (Pt.II), 1958	106 151		1	
Kraus,E.H., Hunt,W.F. & Ramsdell,L.S.	Mineralogy -New York: McGraw Hill, (1)1920, (5)1959	686	13,62	15	U
Krishnan,R.S. (ed.)- see Progress in crystal physics, List III					
Krivoglaz,M.A. & Smirnov,A.A.	Teorija uporjadočivajuščikhsja splavov [Theory of order-disorder in alloys] -Moskva: NAUKA, 1958	388		11,12	
Kröger,F.A.	Chemistry of imperfect crystals -Amsterdam: North Holland, 1964	1039		7,11	

Author	Title, publisher, date	No. of pages	Review ref.	Subject ref.	Te in lev
Kurdjumov,G.V. (ed.)	Rentgenografija v primenenii k issledovani ju materialov [X-rays applied to the study of materials] -Moskva: GNTI, 1936	567		5	
Kurdjumov,G.V.	Javlenija zakalki i otpuska stali [Phenomena of inclusions and voids in steel] -Moskva: Metallurgija, 1960	64		13,12	
Kuznecov,V.D.	Kristally i kristallizacija [Crystals and crystallisation] -Moskva: Gostekhizdat., 1953	411		13	
Kuznecov,V.D.	Poverkhnostaja ênergija tverdykh tel [Surface energy of solids] -Moskva: Gostekhizdat., 1954	220		10	
[Kuznecov,V.D.=] Kutznetsov,V.D.	(tr. from Russian, 1954) Surface energy of solids [D.S.I.R. translation] -London: H.M.S.O., 1957	283		10	
[Kuznecov,V.D.=] Kusnezow,W.D.	(ed. Rackow,B.; tr. Urban,K. from Russian, 1954) Einfluss der Oberflächenergie auf das Verhalten fester Körper -Berlin: Akademie Verlag, 1961	253		10	
Landolt-Börnstein	Zahlenwerte und Funktionen - see List III				
Lapadu-Hargues,P.	Précis de minéralogie -Paris: Masson, 1954	310	8,69	15	
Laue,M.von	Röntgenstrahlinterferenzen -Leipzig: Akademische Verlagsg., (1)1941, (2)1948	409		4	
Laue,M.von	Materiewellen und ihre Interferenzen -Leipzig: Akademische Verlagsg., (1)1944, (2)1948	392	3,403	4	
Laue,M.von	Röntgenwellenfelder in Kristallen -Berlin: Akademie Verlag, 1959	26		4	
Laue,M.von	(ed. Rinne,F. & Schiebold,E.) Die Interferenz d. Röntgenstrahlen (1912-14) -Leipzig: Akademische Verlagsg., 1923	111		4,20	
Laue,M.von	(ed. Kohler,M.) Ausgewählte Schriften und Vorträge -Braunschweig: Vieweg, 1961 Bd.1 Bd.2 Bd.3	 548 513 265	15,517	20	
Laves,F. & Fuster,J.M. (eds.)	Felspars - see Conference 1960, List II				
Lawson,W.D. & Nielsen,S.	Preparation of single crystals -London: Butterworths, 1958	255		13	

Main List - I

Author	Title, publisher, date	No. of pages	Review ref.	Subject ref.	Teaching level
Lemmlejn,G.G.	Sektorial'noe stroenie kristallov [Sector structures in crystals] -Moskva: NAUKA, 1948	40		13	
Levenberg,N.V. (ed.)	Sbornik statej posvjaščennykh pamjati prof.Mikheev,V.I. [Commemorative volume for Prof. V.I. Mikheev - collected articles] - see Kristallografija, List III				
Levinson-Lessing,Ju.G. & Beljankin,D.S.	Učebnik kristallografii, č.1: geometričeskaja kristallografija [Textbook of crystallography, Pt.1: geometrical crystallography] Moskva: Gosizdat., 1923	141		2	
Liebisch,T.	Physikalische Kristallographie -Leipzig: Von Veit, 1891	614		9	
Likhtman,V.I., Šukin,E.D. & Rebinder,P.A.	Fiziko-khimičeskaja mekhanika metallov [Physico-chemical mechanics of metals] -Moskva: NAUKA, 1962	303		12,9	
Lipscomb,W.N.	Boron hydrides -New York: Benjamin, 1963	275		8	
Lipson,H. & Cochran,W.	The determination of crystal structures (The crystalline state, Vol.3) -London: Bell, 1953	345	7,525	3	UG
Lipson,H. & Cochran,W.	(tr. & ed. Belov,N.V.) Opredelenie struktury kristallov [The determination of crystal structures] -Moskva: MIR, 1956	415		3	
Lipson,H. & Taylor,C.A.	Fourier transforms and X-ray diffraction -London: Bell, 1958	76	12,481	3	U
Ljubarskij,G.Ja.	Teorija grupp i ee priloženija v fizike [Group theory and its applications in physics] -Moskva: NAUKA, 1958	354		6,9	
Lomont,J.S.	Applications of finite groups -New York: Academic Press, 1959	346	13,61	6	
Lonsdale,K.	Simplified structure factor and electron density formulae for the 230 space groups of mathematical crystallography -London: Bell, 1936	181		T	
Lonsdale,K.	Crystals and X-rays -London: Bell, 1948	199	3,79	1	U
Lonsdale,K.	(tr. Shaskolskaya,M.P.; ed. Belov,N.V., from English, 1948) Kristally i rentgenovskie luči [Crystals and X-rays] -Moskva: MIR, 1952	216		1	

Main List - I

Author	Title, publisher, date	No. of pages	Review ref.	Subject ref.	Tea in lev
Lonsdale,K. (ed.)	- see International tables, List III				
Ložnikova,O.N. & Jakovleva,S.V.	Rentgenometričeskij spravočnik opredelitel' mineralov soderžaščikh redkozemel'nye ělementy [X-ray reference book for the identification of minerals containing rare-earth elements] -Moskva: NAUKA, 1964	223		15,T	
McClure,D.S.	Electronic spectra of molecules and ions in crystals [repr. from Solid state physics, 1959] -New York: Academic Press, 1964	176		19,9	
Machatschki,F.	Grundlagen der allgemeinen Mineralogie und Kristallchemie -Wien: Springer, 1946	209	1,46	15	[illegible]
Machatschki,F.	Spezielle Mineralogie auf geochemischer Grundlage -Wien: Springer, 1953	378	7,523	15	[illegible]
McLachlan,D.	X-ray crystal structure -New York: McGraw Hill, 1957	416	11,756	1	
McWeeny,R.	Symmetry: An introduction to group theory and its applications -Oxford: Pergamon, 1963	248		6	
Makarov,E.S.	Kristallokhimija prostejšikh soedinenij urana, torija, plutonija, neptunija [Crystal chemistry of simple compounds of uranium, thorium, plutonium and neptunium] -Moskva: NAUKA, 1958	142	14,1102	8	
Makarov,E.S.	(tr. Uvarov,E.V., from Russian, 1958) Crystal chemistry of simple compounds of uranium, thorium, plutonium, and neptunium -New York: Consultants Bureau, 1959	145		8	
Mandelcorn,L. (ed.)	Nonstoichiometric compounds -New York: Academic Press, 1964	674		7	
Mandelkern,L.	Crystallisation of polymers -McGraw Hill, 1964	359		17,13	
Maradudin,A.A., Montroll,E.W. & Weiss,G.H.	Theory of lattice dynamics in the harmonic approximation [Solid state physics, Supplement 3] -New York: Academic Press, 1963	319		9	
Mariot,L.	Groupes finis de symétrie et recherche de solutions de l'Équation de Schroedinger -Paris: Dunod, 1959			6	
Mariot,L.	(tr. Nussbaum,A., from French, 1959) Group theory and solid state physics -London: Prentice Hall, 1963	128		6	
Mark,H.	Die Verwendung der Roentgenstrahlen in Chemie und Technik -Leipzig: Barth, 1926	528		5	

Main List - I

Author	Title, publisher, date	No. of pages	Review ref.	Subject ref.	Teaching level
Mark,H. & Wierl,P.	(tr.: ed. Lebedev,A.A., from German) Difrakcia ėlektronov [Electron diffraction] -Moskva: Gostekhizdat., 1933	189		5	
Matossi,F.	Gruppentheorie der Eigenschwingungen von Punktsystem -Berlin: Springer, 1961	191		6	G
Mauguin,C.	La structure des cristaux determinée au moyen des rayons X -Paris: Blanchard, 1924	281		1	U
Megaw,H.D.	Ferroelectricity in crystals -London: Methuen, 1957	220	11,754	9,7	UG
Melville,H.	Big molecules -London: Bell, 1958	180		17	E
Meyer,K.H. & Mark,H.	Aufbau der hochpolymeren organischen Naturstoffe -Leipzig: Akademische Verlagsg., 1930			17	
Mikheev,V.I.	Rentgenometričeskij opredelitel' mineralov [X-ray handbook of minerals] -Moskva: NEDRA, 1957	868		15	
Mikheev,V.I.	Gomologija kristallov [Homologous crystals] -Leningrad: Gostoptekhizd., 1961	208			
Mirkin,L.I.	(ed. Umanskij,Ja.S.) Spravočnik po rentgenostrukturnomu analizu polukristallov [Collection of data on X-ray analysis of polycrystals] -Moskva: NAUKA, 1961	863		T,5	
Mirkin,L.I.	(tr. from Russian, 1961) Handbook of X-ray structure analysis of polycrystalline materials -New York: Consultants Bureau, 1964			T,5	
Miyake,S. (ed.)	Magnetism and crystallography - see Conference 1961, List II				
Morse,J.K.	Bibliography of crystal structure (1912-1927) -Chicago: Univ.Press, 1928	164		20,T	
Moss,T.S.	Optical properties of semi-conductors -London: Butterworths, 1959	279		14,9	
Mott,N.F.	Atomic structure and the strength of metals -London: Pergamon, 1956	64		12	E
Mott,N.F.	(tr. Guéron,G., from English, 1956) La structure atomique et la résistance des métaux -Paris: Dunod, 1958	102	11,756	12	E
Mott,N.F.	(tr. Träuble,H., from English, 1956) Atomare Struktur und Festigkeit der Metalle -Braunschweig: Vieweg, 1961	56		12	E

Main List - I

Author	Title, publisher, date	No. of pages	Review ref.	Subject ref.	Te i le
Mott,N.F. & Gurney,R.W.	Electronic processes in ionic crystals -Oxford: Univ.Press, (1)1940, (2)1948, 1957 -New York: Dover, repr. 1964	275		9	
Mott,N.F. & Jones,H.	The theory and the properties of metals and alloys -Oxford: Univ.Press, (1)1936, repr. 1962	326		12,9	
Möllestedt,G. & others (eds.)	Elektronenmikroskopie - see Conference 1958, List II				
Mueller,W.M. (ed.)	- see Advances in X-ray analysis, List III				
Mullin,J.W.	Crystallisation -London: Butterworths, 1961	268		13	
Muto,T., & Takagi,Y.; Guttman,L.	The theory of order-disorder transitions in alloys; & Order-disorder phenomena in metals [repr. from Solid state physics 1955/6] -New York: Academic Press, 1964	170		11,12,16	
Nakaya,U.	Snow Crystals -Cambridge (Mass.): Harvard Univ.Press, 1954	510		13	
Neff,H.	Grundlagen und Anwendung der Röntgen-Feinstruktur-Analyse -München: Oldenbourg, (1)1959, (2)1962	460		1	
Neugebauer,C.A., Newkirk,J.B. & Vermilyea,D.A. (eds.)	Structure and properties of thin films - see Conference 1959, List II				
Newkirk,J.E. & Wernick,J.H. (eds.)	Direct observation of imperfections in crystals - see Conference 1961, List II				
Niggli,P.	Geometrische Kristallographie des Diskontinuums -Berlin: Borntraeger, 1919	587		2	
Niggli,P.	Kristallographische und strukturtheoretische Grundbegriffe -Leipzig: Akademische Verlagsg., 1928	317		1	
Niggli,P.	Lehrbuch der Mineralogie und Kristallchemie -Berlin: Borntraeger, Teil 1, 1941 Teil 2, 1942	688 224		15	
Niggli,P.	Grundlagen der Stereochemie -Basel: Birkhäuser, 1945	283		7,1	
Niggli,P. (with Brandenberger,E. & Nowacki,W.)	(tr. & ed. Bokij,G.B., from German) Stereokhimija. Stereokhimičeskaja klassifikacija; 50 let obščej teorii struktury kristallov; simmetrija i fiziko-khimičeskie svojstva kristalličeskikh soedinenij [Stereochemistry. Stereochemical (cont.)				

Author	Title, publisher, date	No. of pages	Review ref.	Subject ref.	Teaching level
	classification; 50 years of theories of crystal structure; symmetry and physical and chemical properties] -Moskva: MIR, 1949	364		7,1	
Niggli,P.	Probleme der Naturwissenschaften -Basel: Birkhäuser, 1949	280	3,484	21	EU
Nowacki,W.	Moderne allgemeine Mineralogie (Kristallographie) -Braunschweig: Vieweg, 1951	54	5,860	15	U
Nowacki,W.	Fouriersynthese von Kristallen und ihre Anwendung in der Chemie -Basel: Birkhäuser, 1952	237	6,224	3,7	UG
Nyburg,S.C.	X-ray analysis of organic structures -New York: Academic Press, 1961	434	15,624	3,7	U
Nye,J.F.	Physical properties of crystals -Oxford: Univ.Press, 1957	322	11,666	9	UG
O'Daniel,H. (ed.)	Zur Struktur und Materie der Festkörper - see Conference 1951, List II				
Oelsner,H.O.	Atlas der wichtigsten Mineralparagenesen in mikroskopischen Bild -Freiberg: Bergakademie, 1961	309		15	
Orgel,L.E.	Introduction to transition-metal chemistry: ligand-field theory -London: Methuen, 1960	180		7	U
Ormont,B.F.	Struktura neorganičeskikh veščestv [Structure of inorganic substances] -Moskva: Gostekhizdat., 1950	968		7	
Padurov,N.N.	Kristallokhimičeskij analiz i metody geometričeskoj kristallografii [Crystal chemical analysis and methods of geometrical crystallography] -Moskva: Gostekhizdat., 1931	272		1	
Palache,C., Berman,H. & Frondel,C.	- see Dana's system of mineralogy, List III				
Parrish,W. (ed.)	Advances in X-ray diffractometry and X-ray spectrography -Eindhoven: Centrex, 1962	249		5	
Parrish,W. (ed.)	World directory of crystallographers - see Smits,D.W.				
Parrish,W., Ekstein,M.G. & Irwin,B.W.	Data for X-ray analysis -Eindhoven: Philips Technical Press, (1)1953		6,877	T,5	
	Vol.I. Charts for solution of Bragg's equation [K wavelengths of Mo,Cu,Co, Fe,Cr]	108			
	Vol.II. Tables for computing the lattice constants of cubic crystals	90			

Main List - I

Author	Title, publisher, date	No. of pages	Review ref.	Subject ref.	Tea in lev
Parrish,W. & Mack,M.	Data for X-ray analysis -Eindhoven: Philips Technical Press, (2)1963 Vols.I-III. Charts for the solution of Bragg's equation I - Cu K radiation II - Mo K, Co K, W L radiations III - Fe K, Cr K radiations Vol.IV. Tables for computing lattice parameters Vol.V. Reflection angle tables for calibration standards.	 125 141 137		T,5	
Partington	- see List III				
Pattee,H.H., Cosslett,V.E. & Engström,A. (eds.)	X-ray optics and X-ray microanalysis - see Conference 1962, List II				
Paul,W. & Warschauer,D.M. (eds.)	Solids under pressure [Bridgman memorial collection] -New York: McGraw Hill, 1963	478		9	
Pauling,L.	The nature of the chemical bond -Ithaca,N.Y.: Cornell Univ.Press, (1)1939, (3)1960	644		7	U
Pearson,W.B.	A handbook of lattice spacings of metals and alloys -London: Pergamon, 1958	1044	12,174	T,12	
Pearson,W.B. (ed.)	- see Structure Reports, List III				
Peierls,R.E.	Quantum theory of solids -Oxford: Univ.Press, 1955	229		9	
Peiser,H.S., Rooksby,H.P. & Wilson,A.J.C.	X-ray diffraction by polycrystalline materials -London: Institute of Physics, (1)1955, corr. 1960	724	8,856 14,696	5	
Pekar,S.I.	Issledovanija po èlektronnoj teorii kristallov [Investigations in the electron theory of crystals] -Moskva: Gostekhizdat., 1951	256	9,94	9	
Pekar,S.I.	(tr. Vogel,H., from Russian, 1951) Untersuchungen über die Elektronentheorie der Kristalle -Berlin: Akademie-Verlag, 1954	184	9,94	9	
Penkovskii,V.V.	Deistve oblučenija na metally i nekotorye tugoplavkie materialy [Effect of radiation on metals and other high-melting materials] -Kiev: Akad.nauk.Ukr.SSR, 1962			11,9	
Penkovskii,V.V.	(tr. from Russian, 1962) Effect of radiation on metals and other high-melting materials -Amsterdam: Elsevier, 1964	201		11,9	

Main List - I

Author	Title, publisher, date	No. of pages	Review ref.	Subject ref.	Teaching level
Pepinsky,R. (ed.)	Computing methods and the phase problem in X-ray crystal analysis - see Conference 1952, List II				
Pepinsky,R., Robertson,J.M. & Speakman,J.C. (eds.)	Computing methods and the phase problem in X-ray crystal analysis - see Conference 1960, List II				
Perutz,M.F.	Proteins and nucleic acids [Weizman memorial lectures] -Amsterdam: Elsevier, 1962	211	17,73	17	
Pfeiffer,H.	Wunderwelt der Kristalle -Leipzig: VEB Deutscher Verlag für Grundstoffindustrie, 1962	47		1	E
Philipsborn,H.von	Tafeln zum Bestimmen der Minerale nach äusseren Kennzeichen -Stuttgart: Schweizerbart, 1953	244	7,383	T,15	
Phillips,F.C.	An introduction to crystallography -London: Longmans, (1)1946, (3)1963	340	1,228	2	U
Physical Society, London	Strength of solids - see Conference 1947, List II				
Physical Society, London	Defects in crystalline solids - see Conference 1955, List II				
Physical Society of Japan	Magnetism and crystallography - see Conference 1961, List II				
Physical Society of Japan	Crystal lattice defects - see Conference 1962, List II				
Picon,M. & Flahaut,J.	Elements de minéralogie et de cristallographie -Paris: Sedes, 1957	292		15	
Pines,B.Ja.	Lekcii po strukturnomu analizu [Lectures on structure analysis] -Kharkov: Gostekhizdat. Ukrainy, (1)1937, (2)1957	455		3	
Pines,B.Ja.	Ostrofokusnye rentgenovskie trubki i priklad rentgenostrukturnyi analiz [Microfocus X-ray tubes and their applications to X-ray structure analysis] -Moskva: Gostekhizdat., 1955	267		5	
Pines,D.	Elementary excitations in solids -New York: Benjamin, 1963	299		9,6	
Pinsker,Z.G.	Difrakcija èlektronov [Electron diffraction] -Moskva: NAUKA, 1949	404		5	
Pinsker,Z.G.	(tr. Spink,J.A. & Feigl,E., from Russian, 1949) Electron diffraction -London: Butterworths, 1953	450		5	

Main List - I

Author	Title, publisher, date	No. of pages	Review ref.	Subject ref.	Te... i... le...
Pirenne,M.H.	The diffraction of X-rays and electrons by free molecules -Cambridge (England): Univ.Press, 1946	166		4	
Pockels,F.	Lehrbuch der Kristalloptik -Leipzig: Teubner, 1906	519		14	
Popov,G.M. & Shafranovskii,I.I.	Kristallografija [Crystallography] -Moskva: GNTI, (1)1941, (3)1955	294		1	
Porai-Koshits,M.A.	[See also Bokij,G.B. & Porai-Koshits,M.A.] Praktičeskii kurs rentgenostrukturnogo analiza, Tom II [Practical course of X-ray analysis, Vol.II] -Moskva: Izdat.Univ., 1960	631		1	
Porai-Koshits,E.A. (ed.)	- see Glassy state (Vol.3), List III				
Porter,M.W. & Spiller,R.C.	- see Barker index of crystals, List III				
Post,E.J.	Formal structure of electromagnetics -Amsterdam: North Holland, 1962	204		9	
Przibram,K.	Verfärbung und Lumineszenz: Beiträge zur Mineralphysik -Wien: Springer, 1953	275	7,382	9,15	
Przibram,K.	(tr. & rev. Caffyn,J.E., from German, 1953) Irradiation colours and luminescence -London: Pergamon, 1956	332	9,840	9,15	
Raaz,F. & Kohler,A.	Bau und Bildung der Kristalle -Wien: Springer, 1953	185	7,524	1,7	
Raaz,F. & Tertsch,H.	Geometrische und physikalische Kristallographie -Wien: Springer, (1)1939, (3)1958	367	5,298 11,667	1	
Rado,G.T. & Suhl,H. (eds.)	Magnetism: a treatise on modern theory and materials. Vol.3, Spin arrangements and crystal structure, domains, and micromagnetics -New York: Academic Press, 1963	620		9	
Raimes,S.	The wave mechanics of electrons in metals -Amsterdam: North Holland, 1961	382		9,12	
Ramachandran,G.N. (ed.)	Advanced methods of crystallography - see Conference 1963, List II				
Ramachandran,G.N. (ed.)	Aspects of protein structure - see Conference 1963, List II				
Ramachandran,G.N. (ed.)	Crystallography and crystal perfection - see Conference 1963, List II				

Main List - I

Author	Title, publisher, date	No. of pages	Review ref.	Subject ref.	Teaching level
Ramdohr,P.	Klockmann's Lehrbuch der Mineralogie -Stuttgart: Enke, (14)1954	674		15	U
Rand,M.H. (ed.)	The measurement and interpretation of diffraction line profiles - see Conference 1961, List II				
Randall,J.T.	The diffraction of X-rays and electrons by amorphous solids, liquids, and gases -London: Chapman & Hall, 1934	290		18,5	
Rassweiler,G.M. & Grube,W.L. (eds.)	Internal stresses and fatigue in metals - see Conference 1958, List II				
Read,W.T.	Dislocations in crystals -New York: McGraw Hill, 1953	228	7,522	11	
Read,W.T.	(tr. Coulomb,P., from English, 1953) Les dislocations dans les cristaux -Paris: Dunod, 1957	237	10,390	11	
Rees,A.L.G.	Chemistry of the defect solid state -London: Methuen, 1954	136		7,9	
Rees,A.L.G.	(tr. from English, 1954) Khimija kristallov s defektami [Chemistry of the defect solid state] -Moskva: MIR, 1956	135		7,9	
Reser,M.K., Smith,G. & Insley,H. (eds.)	Nucleation and crystallisation in glasses and melts - see Conference 1961, List II				
Rinne,F. & Berek,M.	(ed. Claussen,C.H. & others) Anleitung zu optischen Untersuchungen mit dem Polarisationsmikroskop -Stuttgart: Schweizerbart, (1)1933, (2)1953	366	7,382	14	
Rinne,F. & Schiebold,E.	Zur Nomenklatur der 32 Kristallklassen. Über eine neue Herleitung und Nomenklatur der 230 kristallographischen Raumgruppen [Reprint from Vol.90 of Abh. der Math. Phys. Klasse der Sächsischen Akad. der Wissens.; with 18 p. of 4-colour diagrams] -Leipzig: Hirzel, 1929	204		2	U
Rinne,F. & Schiebold,E. (eds.)	- see Laue,M.von				
Robertson,J.M.	Organic crystals and molecules -Ithaca,N.Y.: Cornell Univ.Press, 1953	340	7,523	3,7	U
Rodewald,H.J.	Zur Genesis des Diamanten -Schaffhausen (Schweiz): Meier, 1960	69		8,15	
Romanov,E.M.	Geometričeskaja kristallografia. S pril. nomenklatury avtora dlja sistem klassov i form [Geometrical crystallography, with appendix giving author's nomenclature for systems of classes and forms] -Baku: Akad. nauk. ISSR, 1932	264		2	

Main List - I

Author	Title, publisher, date	No. of pages	Review ref.	Subject ref.	Tea in lev
Rose,A.J.	Tables permettant le dépouillement des diagrammes de rayons X et abaques de réglage des monochromateurs à lame courbe -Paris: C.N.R.S., 1957	141	11,900	T,5	
Rose,A.J. (ed.)	Index of manufacturers of apparatus and materials used in crystallography -International Union of Crystallography, (2)1959	126		T	
Rubber Handbook	- see Hodgman,C.D. (ed.)				
Sachs,M.	Solid state theory -New York: McGraw Hill, 1963	360		9	
Sachse,H.	Ferroelektrika -Berlin: Springer, 1956	171		9	
Sachse,H.	(tr. Bonnet,A., from German, 1956) Les ferroélectriques -Paris: Dunod, 1958	186		9	
Sagel,K.	Tabellen zur Röntgenstrukturanalyse -Berlin: Springer, 1958	204	13,854	T	
Sagel,K.	Tabellen zur Röntgen-emissions und Absorptionsanalyse, 1959	135		T	
Samojlov,Ja.V.	Vvedenie b kristallografiju [Introduction to crystallography] -Moskva: NEDRA, 1932	128		1	
Sanderson,R.T.	Teaching chemistry with models -Princeton (N.J.): Van Nostrand, 1962	175		7	
Sarančina,G.M.	Fedorovskij metod [The Fedorov method] -Leningrad: Izdat.Univ., (1)1954, (2)1963	153		14	
[Sarančina,G.M.=]	Sarantschina,G.M. (tr. from Russian) Die Fedorow-Methode -Berlin: VEB Deutscher Verlag der Wissenschaften	135		14	
Saratovkin,D.D.	Dendritnaja kristallizacia [Dendritic crystallisation] -Moskva: Metallurgija, (2)1957	127		13	
Saratovkin,D.D.	(tr. Bradley,J.E.S., from Russian) Dendritic crystallisation -New York: Consultants Bureau, 1959	126	13,364	13	
Savitsky,E.M.	Vlijanie temperatury na mekhaničeskie svojstva metallov i splavov [The influence of temperature on the mechanical properties of metals and alloys] -Moskva: NAUKA, 1957	294		9,12	
Savitsky,E.M.	(ed. Sherby,O.D., tr. Sherby,D., from Russian, 1957) The influence of temperature on the mechanical properties of metals and alloys -Stanford (Cal.): Univ.Press, 1961	303		9,12	

Main List - I

Author	Title, publisher, date	No. of pages	Review ref.	Subject ref.	Teaching level
Schiebold,E.	Methoden der Kristallstrukturbestimmung mit Röntgenstrahlen, Bd.I - Die Lauemethode -Leipzig: Akademische Verlagsg., 1932	173		5	U
Schieltz,N.C.	The application of the reciprocal lattice concept in the graphical solution of X-ray diffraction problems -Golden (Col.): Colorado School of Mines, 1964	70		5,2	U
Schmid,E. & Boas,W.	Kristallplastizität, mit besonderer Berücksichtigung der Metalle -Berlin: Springer, 1935	373	4,384	9,12	UG
Schmid,E. & Boas,W.	(tr. from German) Plasticity of crystals -London: Hughes, 1950	353	4,384	9,12	
Schoenflies,A.	Kristallsysteme und Krystallstruktur -Leipzig: Teubner, 1891	638		2	G
Schoenflies,A.	Theorie der Kristallstruktur -Berlin: Borntraeger, 1923	567		2	G
Schroeder,R.	Kristallometrisches Praktikum -Berlin: Springer, 1959	119	4,576	2	U
Schubert,K.	Kristallstrukturen zweikomponentiger Phasen -Berlin: Springer, 1964	432		7,T	
Seel,F.	Atombau und chemische Bindung -Stuttgart: Enke, (1)1956, (3)1960	96		7	U
Seel,F.	(tr. & rev. Greenwood,N.N. & Stadler,H.P., from German 4th ed.) Atomic structure and chemical bonding -London: Methuen, 1963	120		7	
Segal,J., Dornberger-Schiff,K. & Kalaidjiev,A.	(tr. Wooster,A.) Globular protein molecules, their structure and dynamic properties -Berlin: VEB Deutscher Verlag der Wissenschaften, 1961	150	17,75	17	
Seitz,F. & Turnbull,D. (eds.)	- see Solid state physics, List III				
Shafranovskii,I.I.	Kristallografija okruglykh almazov [Crystallography of rounded diamonds] -Leningrad: Izdat.Univ., 1948	132		13,8	
Shafranovskii,I.I.	Kristally mineralov, č.1: ploskogrannye formy [Crystals of minerals, Pt.1: plane-bounded forms] -Leningrad: Izdat.Univ., 1957	222		15,13	
Shafranovskii,I.I.	Kristally mineralov: krivogrannye, skeletnye i zernistye formy [Crystals of minerals: rounded, skeletal and granular forms] -Moskva: NEDRA, 1961	332		15,13	

Main List - I

Author	Title, publisher, date	No. of pages	Review ref.	Subject ref.	Tes ir lev
Shafranovskii,I.I.	Lekcii po kristallomorfologii mineralov [Lectures on the morphology of mineral crystals] -Lvov: Izdat.Univ., 1960	162		15,13	
Shafranovskii,I.I.	Istorija kristallografii v Rossii [History of crystallography in Russia] -Leningrad: NAUKA, 1962	415		20	
Shafranovskii,I.I.	Evgraf Stepanovich Federov -Moskva: NAUKA, 1963	283		20	
Shafranovskii,I.I.	& Mikheev,V.I. (eds.) - <u>see</u> Kristallografija, List III				
Shaskolskaya,M.P.	Kristally [Crystals] -Moskva: Gostekhizdat., 1956	228		1	I
[Shaskolskaya,M.P.=]	Chaskolskaia,M. (tr. Rygalov,M. from Russian) Les cristaux -Moskva: MIR, 1959	287		1	I
Shell Petroleum	X-ray diffraction patterns of lead compounds -Chester (England): Shell Petroleum, 1954	81	<u>8</u>,69	T,5	
Shockley,W.	Electrons and holes in semi-conductors -New York: Van Nostrand, 1950	592		9	
Shockley,W., Hollomon,J.H., Maurer,R. & Seitz,F.	Imperfections in nearly perfect crystals - <u>see</u> Conference 1950, List II				
Shubnikov,A.V.	Simmetrija: Zakony simmetrii i ikh primenenie v nauke, tekhnike i prikl. iskusstve [Symmetry: the laws and their application to science, technology and applied art] -Moskva: NAUKA, 1940	176		2	
Shubnikov,A.V.	P'ezoelěktričeskie tekstury [Piezoelectric textures] -Moskva: NAUKA, 1946	100		9	
Shubnikov,A.V.	Atlas kristallografičeskikh grupp simmetrii [Atlas of crystallographic symmetry groups] -Moskva: NAUKA, 1946	56		2,T	
Shubnikov,A.V.	Obrazovanie kristallov [Crystal growth] -Moskva: NAUKA, 1947	74		13	
Shubnikov,A.V.	Optičeskaja kristallografija [Optical crystallography] -Moskva: NAUKA, 1950	275		14	
Shubnikov,A.V.	Simmetrija i antisimmetrija konečnykh figur [Symmetry and antisymmetry of finite figures] -Moskva: NAUKA, 1951	172		2	

Main List - I

Author	Title, publisher, date	No. of pages	Review ref.	Subject ref.	Teaching level
Shubnikov,A.V.	Kristally v nauke i tekhnike [Crystals in science and technology] -Moskva: NAUKA, (1)1956, (2)1958	56		1	E
Shubnikov,A.V.	Osnovy optičeskoi kristallografii [The principles of optical crystallography] -Moskva: NAUKA, 1958	205		14	
Shubnikov,A.V.	(tr. from Russian, 1958) Principles of optical crystallography -New York: Consultants Bureau, 1960	186		14	U
Shubnikov,A.V.	Problema dissimmetrii materialn'ykh ob'ektov [Problem of asymmetry of material objects] -Moskva: NAUKA, 1961	75		2	
Shubnikov,A.V., Belov,N.V. & others	(tr. Itzkoff,J. & Gollog,J.; ed. Holser, W.T.; from Russian, 1951-8) Coloured symmetry Oxford: Pergamon, 1964	263		2	
Shubnikov,A.V., Flint,E.E. & Bokij,G.B.	Osnovy kristallografii [Principles of crystallography] -Moskva: NAUKA, 1940	488		1	
Shubnikov,A.V. & Sheftal,N.N. (eds.)	- see Rost kristallov [Growth of crystals] List III				
Shubnikov,A.V. & Sheftal,N.N. (eds.; tr. from Russian)	Growth of crystals - see List III, entry following Rost kristallov				
Shubnikov,A.V., Zheludev,I.S., Konstantinova,V.P. & Silvestrova,I.M.	Issledovanie p'ezoėlektričeskikh tekstur [Investigation of piezoelectric textures] -Moskva: NAUKA, 1955	189		9	
Shubnikov,A.V., Zheludev,I.S., Konstantinova,V.P. & Silverstrova,I.M.	(tr. Daknoff,A., from Russian) Études des textures piézoélectriques -Paris: Dunod, 1958	208		9	
Sirota,N.N. (ed.)	Kristallizačija i fazovye perekhody [Crystallisation and phase changes] -Minsk: Akad.nauk.BSSR, 1962	444		16,13	
Sirota,N.N., Belov,K.P. & others (eds.)	Ferrity [Ferrites] - see Conference 1960, List II				
Skobel'cin,D.B. (ed.)	Optičeskie metody issledovanija struktury tverdogo tela [Optical methods of investigation of the structure of solids] -Moskva: NAUKA, 1964	223		14	
Slater,J.C.	Quantum theory of molecules and solids, Vol.1: Electronic structure of molecules -New York: McGraw Hill, 1963	485		7	

Main List - I

Author	Title, publisher, date	No. of pages	Review ref.	Subject ref.	Tea[illegible] in lev[illegible]
Smakula,A.	Einkristalle: Wachstum, Herstellung und Anwendung -Berlin: Springer, 1962	431		13	[illegible]
Smallman,R.E.	Modern physical metallurgy -London: Butterworths, 1962	216		12	[illegible]
Smidt,J. (ed.)	Magnetic and electric resonance and relaxation - see Conference 1962, List II				
Smith,R.A.	Semiconductors -Cambridge (England): Univ.Press, 1959	496		9	
Smith,R.A.	Wave mechanics of crystalline solids -London: Chapman & Hall, 1961	473		9	
Smith,R.A. (ed.)	Semiconductors - see Italian Physical Society, List III				
Smithells,C.J. (ed.)	Metals reference book, Vols.I and II -London: Butterworths, (1)1949, (3)1962	1087		T,12	
Smits,D.W. (ed.)	World directory of crystallographers -Utrecht: Oosthoek, (1)[ed. Parrish,W.] 1957, (2)1960	134		T	
Smoluchowski,R., Mayer,J.E. & Weyl,W.A. (eds.)	Phase transformation in solids - see Conference 1948, List II				
Sobolev,V.	Vvedenie v mineralogiju silikatov [Introduction to silicate mineralogy] -Lvov: Izdat.Univ., 1949	333		15	
Société Française de Minéralogie	Minéralogie-cristallographie, aspects actuels 1878-1953 -Paris: Masson, 1953	1145		15	
Sohncke,L.	Die unbegrenzten regelmässigen Punktsysteme als Grundlage einer Theorie der Kristallstruktur -Karlsruhe: Braun, 1876	83		2	[illegible]
Sohncke,L.	Entwickelung einer Theorie der Kristallstruktur -Leipzig: Teubner, 1879	247		2	[illegible]
Sommerfeldt,E.	Die Kristallgruppen nebst ihren Beziehungen zu den Raumgittern -Dresden: Theodor Steinkopff, 1911	79		2	
Speakman,J.C.	Introduction to electronic theory of valency -London: Arnold, (1)1935, (3)1955	180		7	[illegible]
Speiser,A.	Die Theorie der Gruppen von endlicher Ordnung -Basel: Birkhaüser, (1)1923, (4)1956	272		6	
Spiegel-Adolf,M. & Henny,G.G.	X-ray diffraction studies in biology and medicine -New York: Grune & Stratton, 1947	215		17	
Springall,H.D.	The structural chemistry of proteins -London: Butterworths, 1954	376		17	

Main List - I

Author	Title, publisher, date	No. of pages	Review ref.	Subject ref.	Teaching level
Stadelmaier,H.H. & Austin,W.W. (eds.)	Materials Science Vol.1 - see Conference 1962, List II				
Standley,K.J.	Oxide magnetic materials -Oxford: Univ.Press, 1962	160	15,922	8,9	
Stasiw,O.	Elektronen- und Ionenprozesse in Ionenkristallen -Berlin: Springer, 1959	307		9	G
Steno,N.	(tr.; ed. Belusov,V.V. & Shafranovskii, I.I., from Latin of 1669) O tverdom, estestvenno soderžaščemsja v tverdom [De solido intra solidum naturaliter contento dissertationis prodromus] -Moskva: NAUKA, 1957	151		20	
Stickland,A.C. (ed.)	Physics of semiconductors - see Conference 1962, List II				
Stickland,A.C. (ed.)	- see Reports on progress in physics, List III				
Stokes,A.R.	The theory of the optical properties of inhomogeneous materials -London: Spon, 1963	95		14,11	
Straumanis,M. & Ievins,A.F.	Die Präzisionsbestimmung von Gitterkonstanten nach der asymmetrischen Methode -Berlin: Springer, 1940 -New York: Edwards (reprint) 1948	104		5	
Strunz,H.	Mineralogische Tabellen -Leipzig: Akademische Verlagsg., (1)1941 (3)1957	448	4,79 11,668	15,7,T	
Stumpf,H.	Quantentheorie der Ionenrealkristalle -Berlin: Springer, 1961	278		9	G
Sutton,L.E. (ed.)(with Mitchell,A.D. & others)	Tables of interatomic distances and configurations in molecules and ions [Special publication no.11 of the Chemical Society] -London: Chemical Society, 1958	377	12,174	T,7	
Swalin,R.A.	Thermodynamics of solids -New York: Wiley, 1962	343		9	
Swineford,A. (ed.)	- see Clays and clay minerals, List III				
Šafranovskii	- see Shafranovskii				
Šaskolskaja	- see Shaskolskaya				
Šubnikov	- see Shubnikov				
Taylor,A.	An introduction to X-ray metallography -London: Chapman & Hall, 1945, corr.1952	400		12,5	
Taylor,A.	X-ray metallography -New York: Wiley, 1961	993	15,813	12,5	

Main List - I

Author	Title, publisher, date	No. of pages	Review ref.	Subject ref.	Te i le
Taylor,A. & Kagle,B.J.	Crystallographic data on metal and alloy structures -New York: Dover, 1963	263		T,12	
Taylor,C.A. & Lipson,H.	Optical transforms -London: Bell, 1964	182		3	
Taylor,H.F.W. (ed.)	The chemistry of cements, Vols.1 and 2 -New York: Academic Press, 1964	913		8,15	
Temperley,H.N.V.	Changes of state -London: Macmillan, 1956	324		16	
Terpstra,P.	Kristallometrie -Groningen: Wolters, 1946	372	1,344	2	
Terpstra,P.	A thousand and one questions on crystallographic problems -Groningen: Wolters, 1952	195	6,432	2	
Terpstra,P.	Introduction to the space groups -Groningen: Wolters, 1955	160		2	
Terpstra,P.	Note on the mathematical background of repetitive patterns (translation of introductory chapter to Escher, Grafiek en tekeningen) -London: Oldbourne, 1961	8		21,2	
Terpstra,P. & Codd, L.W.	(tr. from Dutch, 1946) Crystallometry -London: Longmans, 1961	420	16,78	2	
Tertsch,H.	Die Festigkeitserscheinungen der Kristalle -Wien: Springer, 1949	310		9	
Tertsch,H.	Die Stereographische Projektion in der Kristallkunde -Wiesbaden: Verlag für angewandte Wissenschaften, 1954	122	7,856	2	
Thewlis,J. (ed.)	- see Encyclopedic dictionary of physics, List III				
Thibaud,J.	Les rayons X -Paris: Colin, (1)1930, (5)1960	220		1	
Thomas,G.	Transmission electron microscopy for metals -New York: Wiley, 1962	299	15,1064	19,12	
Thomas,G. & Washburn,J.W. (eds.)	Electron microscopy and strength of crystals - see Conference 1961, List II				
Thompson,D.W.	On growth and form -Cambridge (England): Univ.Press, (1)1917, (2)1959 (2 vols.)	1163		21	
Thompson,D.W.	(abridged by Bonner,J.T.) On growth and form Cambridge (England): Univ.Press, 1961	346		21	

Main List - I

Author	Title, publisher, date	No. of pages	Review ref.	Subject ref.	Teaching level
Timmermans, J.	Les constantes physiques des composés organiques cristallines -Paris: Masson, 1953	556	7,606	T,7	
Timmermans, J.	(tr. from French) Physico-chemical constants of pure organic compounds -Amsterdam: Elsevier Vol.1. 1950, repr.1961 Vol.2. (Supplement)	693 450		T,7	
Tolansky, S.	Microstructures of diamond surfaces -London: N.A.G. Press, 1955	67	9,324	10,8	
Tolansky, S.	Surface microtopography -London: Longmans, 1960	304		10	
Tolansky, S.	The history and use of the diamond -London: Methuen, 1962	166		15,21	E
Tolkačev, S.S.	Tablicy mezploskostnykh rasstojanij [Tables of interplanar spacings] -Leningrad: Izdat.Univ., 1955	145		T,5	
Trapeznikov, A.K.	Osnovy rentgenografii [Principles of Röntgenography] -Moskva: Gostekhizdat., 1933	188		1	
Trey, F. & Legat, W.	Einführung in die Untersuchung der Kristallgitter mit Röntgenstrahlen -Wien: Springer, 1954	113		1	U
Trillat, J.J.	Les applications des rayons X - physique, chimie, métallurgie -Paris: P.U.F., 1930	298		1	
Trillat, J.J.	Découverte de la matière -Paris: Albin Michel, 1956	318		1	E
Trillat, J.J.	(tr. from French, 1956) Exploring the structure of matter -New York: Interscience, 1959	214		1	E
Tunell, G. & Murdoch, J.	Introduction to crystallography: a laboratory manual for students of mineralogy and geology -San Francisco: Freeman, (1)1957, (2)1964	55	10,720	1	
Tutton, A.E.H.	The natural history of crystals -London: Kegan Paul, 1924	287		1	EU
Tutton, A.E.H.	Crystalline form and chemical constitution -London: Macmillan, 1926	252		2,7	EU
Tutton, A.E.H.	Crystallography and practical crystal measurement -London: Macmillan, (1)1911 (2)2 vols., 1922 -New York: Hafner, repr.1964	946 1445		1,2	
Tzinzerling	- see Cinzerling				
Ubbelohde, A.R. & Lewis, F.A.	Graphite and its crystal compounds -Oxford: Univ.Press, 1960	217		8	

Main List - I

Author	Title, publisher, date	No. of pages	Review ref.	Subject ref.	Te i le
Umanskij,M.M.	Apparatura rentgenostrukturnykh issledovanij [Apparatus for X-ray structural investigations] -Moskva: NAUKA, 1960	348		5	
Umanskij,Ja.S.	Rentgenografija metallov [X-ray investigation of metals] -Moskva: Metallurgija, 1960	448		12,5	
Umanskij,Ja.S., Trapeznikov,A.K. & Kitaigorodskii,A.I.	Rentgenografija [Röntgenography] -Moskva: Mašinostroenie, 1952	310		5	
Vainštein,B.K.	Strukturnaja ėlektronografija [Structure investigations by electron diffraction] -Moskva: NAUKA, 1956	313		5,3	
[Vainštein,B.K.=]	Vainshtein,B.K. (tr. & ed. Feigl,E. & Spink,J.A., from Russian) Structure analysis by electron diffraction -Oxford: Pergamon, 1964	420		5,3	
Vainštein,B.K.	Difrakcija rentgenovykh lučei na černykh molekulakh [Diffraction of X-rays by chain molecules] -Moskva: NAUKA, 1963	371		4,17	
[Vainštein,B.K.=]	Vainshtein,B.K. (tr. from Russian, 1963, rev.) Diffraction of X-rays by chain molecules -Amsterdam: Elsevier, 1964	350		4,17	
Van Hook,A.	Crystallisation: Theory and practice -New York: Reinhold, 1961	325		13	
Van Vlack,L.H.	Elements of materials science (an introductory text for engineering students) -Reading (Mass.): Addison-Wesley, (1)1959, (2)1964	445		9	
Vardanjanc,L.A.	Triadnija teorija dvojnkovykh obrazovanij mineralov [The triad theory of twinning in minerals] -Erevan: Akad.nauk.Armenia, 1950	108		15	
Vardanjanc,L.A.	Teorija fedorovskogo metodo [Theory of the Fedorov method] -Erevan: Akad.nauk.Armenia, 1959	192		14	
Vasil'ev,L.I.	Dislokacii v metallakh i splavakh [Dislocations in metals and alloys] -Leningrad: Izdat."Znanie", 1964	100		11,12	
Verma,A.R.	Crystal growth and dislocations -London: Butterworths, 1953	182	7,224	11,13	
Vernadskii,V.I.	Osnovy kristallografii, č.1 [Fundamentals of crystallography, Pt.1] -Moskva: Izdat.Univ., 1904	346		2	
Vogel,H.J., Bryson,V. & Lampen,J.O. (eds.)	Informational macromolecules - see Conference 1962, List II				

Main List - I

Author	Title, publisher, date	No. of pages	Review ref.	Subject ref.	Teaching level
Voigt,W.	Lehrbuch der Kristallphysik -Leipzig: Teubner, 1910, 1928 -[U.S.A.] repr. 1944	987		9	
Vul'f	- see Wulff				
Wade,F.A. & Mattox,R.B.	Elements of crystallography and mineralogy -New York: Harper, 1960	332		15	E
Wahlstrom,E.E.	Optical crystallography -New York: Wiley, (1)1943, (3)1960	356	5,296 14,696	14	U
Wahlstrom,E.E.	Petrographic mineralogy -New York: Wiley, 1955	408	9,690	15	
Waller,M.D.	Chladni figures (a study in symmetry) -London: Bell, 1961	193	15,921	9,21	
Walton,J.	Physical gemmology -London: Pitman, 1952	304		15	
Walton,J.	Pocket chart of ornamental and gem stones -London: Pitman, 1954	74		15	
Wassermann,G.	Texturen metallischer Werkstoffe -Berlin: Springer, 1939	194		12	
Wassermann,G. & Grewen,J.	Texturen metallischer Werkstoffe -Berlin: Springer, (2)1962	832	16,708	12	
Watanabé,T. & Takéuchi,W.	Scientific information in the fields of crystallography and solid-state physics - see Conference 1961, List II				
Webster,R.	Gems:(their sources, descriptions and identification) Vols.1 and 2 -London: Butterworths, 1962	792		15	
Wells,A.F.	Structural inorganic chemistry -Oxford: Univ.Press, (1)1945, (3)1962	1076	15,921	7	UG
Wells,A.F.	(tr. from English) Stroenie neorganičeskikh veščestv [Structural inorganic chemistry] -Moskva: MIR, 1948	691		7	
Wells,A.F.	The third dimension in chemistry -Oxford: Univ.Press, 1956	143		7	EU
Werner,A.G.	(tr. Carozzi,A.V., from German of 1774) On the external characters of minerals -Urbana (Ill.): Univ. of Illinois Press, 1962	118		15,20	
Westbrook,J.H. (ed.)	Mechanical properties of intermetallic compounds -New York: Wiley, 1960	435		12,9	
Weyl,H.	Symmetry -Princeton (N.J.): Univ.Press, 1951	168		21	

Author	Title, publisher, date	No. of pages	Review ref.	Subject ref.	Te i le
Weyl,H.	(tr. Bechtolsheim,L. from English) Symmetrie -Basel: Birkhäuser, 1955	160		21	
Wheatley,P.	The determination of molecular structure -Oxford: Univ.Press, 1959	263		7,19	
Wilke,K.T.	Methoden der Kristallzüchtung -Berlin: Deutscher Verlag der Wissenschaften, 1962	250		13	
Wilson,A.J.C.	X-ray optics -London: Methuen, (1)1949, (2)1962	147	16,578	4,11	
Wilson,A.J.C.	(tr.; ed. Iveronova,V.I., from English) Optika rentgenovskikh lučej [X-ray optics] -Moskva: MIR, 1951	144		4,11	
Wilson,A.J.C.	Mathematical theory of X-ray powder diffractometry -Eindhoven: Centrex, 1963	128	17,791	4,5	
Wilson,A.J.C.	(tr. from English, 1963) Théorie mathématique de la diffractométrie des poudres aux rayons X -Eindhoven: Centrex, 1964	131		4,5	
Wilson,A.J.C. (ed.)	- see Structure Reports, List III				
Winchell,A.N.	Microscopical characters of artificial inorganic solid substances or artificial minerals -New York: Wiley, 1931	403		T,15	
Winchell,A.N.	The optical properties of organic compounds -New York: Academic Press, (1)1943, (2)1954	487	9,325	T,7	
Winchell,A.N.	Elements of optical mineralogy (an introduction to microscopic petrography) -New York: Wiley				
	Pt.I: Principles and methods, (1)1922 (5)1937	263		14	
	Pt.II(with Winchell,H.): Descriptions of minerals, (1)1927, (4)1951	551	5,856	15	
	Pt.III: Determinative tables, (2)1929	231		T	
Winkler,H.G.F.	Struktur und Eigenschaften der Kristalle -Berlin: Springer, (1)1950, (2)1955	348	4,575	1	
Witzmann,H. (ed.)	- see Dubinin,M.M. & Witzmann,H. (eds.)				
Wohlraube,E.A.	Crystals -Philadelphia (Pa.): Lippincott, 1962	128		1	
Wolf,K.A.	Struktur und physikalisches Verhalten der Kunstoffe -Berlin: Springer, 1962	974	16,233	17	
Wolf,K.L. & Wolff,R.	Symmetrie -Münster-Köln: Böhlau-Verlag, 1956			2,T	
	Bd.1 Textband	139			
	Bd.2 Tafelband	192			

Main List - I

Author	Title, publisher, date	No. of pages	Review ref.	Subject ref.	Teaching level
Wolkenstein,M.W.	Struktur und physikalische Eigenschaften der Moleküle -Leipzig: Teubner, 1960	770		7,9	G
Wood,E.A.	Crystals and light -Princeton (N.J.): Van Nostrand, 1963	160		14,2	U
Wood,E.A.	Crystal orientation manual -New York: Columbia Univ.Press, 1963	75		5	
Woolfson,M.M.	Direct methods in crystallography -Oxford: Univ.Press, 1961	152		3	G
Wooster,W.A.	A text-book of crystal physics -Cambridge (England): Univ.Press, (1)1938, corr.1949	295		9	U
Wooster,W.A.	Experimental crystal physics -Oxford: Univ.Press, 1957	115	12,80	9	U
Wooster,W.A.	(tr. from English) Experimentelle Kristallphysik -Berlin: Deutscher Verlag , 1958	133		9	U
Wooster,W.A.	(tr. Kopcik,V.A.; ed.Shubnikov,A.V.) Praktičeskoe rukovodstvo po kristallofizike [Experimental crystal physics] -Moskva: MIR, 1958	163		9	
Wooster,W.A.	Diffuse X-ray reflections from crystals -Oxford: Univ.Press, 1962	208	16,855	4,11	G
Wooster,W.A.	(tr. Mirkin,L.I.; ed. Ždanov,G.S.; from English) Diffuznoe rassejanie rentgenovskikh lučej v kristallakh [Diffuse X-ray reflections from crystals] -Moskva: MIR, 1963	287		4,11	
Wrinch,D.	Fourier transforms and structure factors -American Society for X-ray and Electron Diffraction: now American Cristallographic Association, 1946	96		3	
Wrinch,D.	Chemical aspects of the structure of small peptides -København: Munksgaard, 1960	194		8,17	
Wulff,G.V.	Kristally, ikh obrazovanie, vid i stroenie [Crystals: growth, form and structure] - (1)1917, (2)1926	166		13	
Wulff,G.V.	Osnovy kristallografii [Principles of crystallography] -Moskva: Gosizdat., (1)1923, (2)1926			2	
Wulff,G.V.	(ed. Mlodzievskii,A.V.) Izbrannye raboty po kristallofiziki i kristallografii [Collected works on crystal physics and crystallography] -Moskva: Gostekhizdat., 1952	344		20,1	

Main List - I

Author	Title, publisher, date	No. of pages	Review ref.	Subject ref.	Te i le
Wulff,G.V. & Shubnikov,A.V.	Kratkij kurs geometričeskoj kristallografii so stereografičeskoj setkoi [A short course of geometrical crystallography with the stereographic net] -Moskva: Gosizdat., 1924	60		2	
Wyckoff,R.W.G.	The analytical expression of the results of the theory of space groups -Washington: Carnegie Institute, (1)1922, (2)1930	239		2	
Wyckoff,R.W.G.	The structure of crystals -American Chemical Society, (1)1924, (2)1931	497		7	
Wyckoff,R.W.G.	Crystal structures -New York: Interscience (1)1948 and onwards [serial publication in loose-leaf form]			7,T	
	(2)Vol.1, 1963	467	7,867		
	(2)Vol.2, 1964	588	15,173		
Zachariasen,W.H.	Theory of X-ray diffraction in crystals -New York: Wiley, 1946	255		4	
Zavarickij,A.N. (ed.)	Universal'nyi stolik E.S. Fedorova [Fedorov's universal stage][collected articles] -Moskva: NAUKA, 1953	838		14	
Ziman,J.M.	Electrons and phonons (the theory of transport phenomena in solids) -Oxford: Univ.Press, 1960	553		9	
Ziman,J.M.	Electrons in metals - a short guide to the Fermi surface -London: Taylor & Francis, 1962	80		9,12	
Ziman,J.M.	Principles of the theory of solids -Cambridge (England): Univ.Press, 1964	360		9	
Zhdanov	- see Ždanov				
Ždanov,G.S.	Osnovy rentgenovskogo strukturnogo analiza [Principles of X-ray structure analysis] -Moskva: Gostekhizdat., 1940	446		1	
Ždanov,G.S.	Fizika tverdogo tela [Solid state physics] -Moskva: Izdat.Univ., 1961	501		9	
Ždanov,G.S. & Umanskij,Ja.S.	Rentgenografija metallov [X-ray investigation of metals] -Moskva: Metallurgija			12,5	
	č.1 (Pt.1) (1)1937, (2)1941	392			
	č.2 (Pt.2) 1938	387			

CONFERENCES - II

Explanation

Conferences are arranged chronologically by the year of their occurrence; and within the same year, alphabetically by the distinctive part of their title. (General phrases such as 'Proceedings of an international symposium on . . ' are either omitted or put in brackets after the distinctive part.) The place where the Conference was held is given immediately after the title.

The name of the editor is given, if known. If no editor's name is given, the name of the scientific society acting as sponsor is recorded instead.

For an explanation of the other information given in this list, see the preface to List I.

Note the following exceptions which have not been included in List II:
(a) Conference reports which have not been recognised as such will be found under their editors' names in List I.
(b) Certain Conference Series, with the same title and the same organisation over a number of years, will be found under their titles in List III. They are as follows:

Advances in X-ray analysis (Denver conferences)
Clays and clay minerals (National Conferences on . . .)
Glassy state (tr. from Russian)
Growth of crystals (tr. from Russian)
Italian Physical Society - International School "Enrico Fermi"
Kristallografija - Trudy fedorovskoj naučnoj sessii
Rost kristallov
Trudy instituta Kristallografii

Year	Title, place, editor, publisher	No. of pages	Review ref.	Subject ref.
1947	Strength of solids: Bristol -London: Physical Society, 1948	162		11,9
1948	Phase transformations in solids: Cornell Univ. Ed. Smoluchowski,R., Mayer,J.E. & Weyl,W.A. -New York: Wiley, 1951, repr.1957	660	5,857	16
1949	Crystal growth: London Faraday Society -London: Butterworths, repr.1959	366		13
1950	Imperfections in nearly perfect crystals: Pocono Manor Ed. Shockley,W., Hollomon,J.H., Maurer,R. & Seitz,F. -New York: Wiley, 1952	490	6,110	11
1951	Zur Struktur und Materie der Festkörper: Frankfurt a.M. Ed. O'Daniel,H. -Berlin: Springer, 1952	304	6,368	13,7
1952	Chemistry of cement (Proceedings of the third international symposium): London -London: Cement & Concrete Association, 1954	870		8
1952	Computing methods and the phase problem in X-ray crystal analysis: Pennsylvania State Univ. Ed. Pepinsky,R. -University Park (Pa.): X-ray Crystal Analysis Laboratory, 1952	390	6,876	3

Conferences - II

Year	Title, place, editor, publisher	No. of pages	Review ref.	Subj. ref.
1952	The chemistry of solids (Proceedings of the international symposium on the reactivity of solids): Gothenburg -Stockholm: Ingeniörsvetenskapsakademien, 1954	1096		16
1952	Structure and properties of solid surfaces: Lake Geneva (Wis.) Ed. Gomer,R. & Smith,C.S. -Chicago: Univ.Press, 1953	491		10
1953	Les applications de la mécanique ondulatoire à l'étude de la structure des molécules: Paris Ed. Broglie,L.de -Paris: Revue d'optique théorique et instrumentale, 1953	223		9
1954	Defects in crystalline solids: Bristol -London: Physical Society, 1955	429	8,855	11
1954	The fibrous proteins: Leeds Ed. Brown,N.R. -Cambridge (England): Univ.Press, 1955	300		17
1956	Actions des rayonnements de grande énergie sur les solides: Paris Ed. Cauchois,Y. -Paris: Gauthier-Villars, 1956	152		11
1956	Dislocations and mechanical properties of crystals: Lake Placid Ed. Fisher,J.C., Johnston,W.G., Thompson,R. & Vreeland,T. -New York: Wiley, 1957	634	11,755	11
1956	La diffusion dans les métaux: Eindhoven Ed. Fast,J.D. -Eindhoven: Philips Technical Press, 1957	134		16
1956	X-ray microscopy and microradiography: Cambridge (England) Ed. Cosslett,V.E., Engström,A. & Pattee,H.H. -New York: Academic Press, 1957	645		19
1958	Diffusion de l'état solide (Colloque sur la ...): Saclay Sponsor: Commissariat à l'énergie atomique -Amsterdam: North Holland, 1959	176		16
1958	Elektronenmikroskopie (Vierter internationaler Kongress): Berlin Band I - Physikalischer-technischer Teil Ed. Möllestedt,G. & others -Berlin: Springer, 1960	851		19
1958	Growth and perfection of crystals: Cooperstown (N.Y.) Ed. Doremus,R.H., Roberts,B.W. & Turnbull,D. -New York: Wiley, 1958	609	12,614	13
1958	Internal stresses and fatigue in metals: Detroit Ed. Rassweiler,G.M. & Grube,W.L. -Amsterdam: Elsevier, 1959	466		11
1958	Non-crystalline solids: Alfred (N.Y.) Ed. Fréchette,V.D. -New York: Wiley, 1960	536		18

Conferences - II

Year	Title, place, editor, publisher	No. of pages	Review ref.	Subject ref.
1959	Structure and properties of thin films: Lake George (N.Y.) Ed. Neugebauer,C.A., Newkirk,J.B. & Vermilyea,D.A. -New York: Wiley, 1959	561		10
1959	X-ray microscopy and X-ray microanalysis: Stockholm Ed. Engström,A., Cosslett,V.E. & Pattee,H.H. -Amsterdam: Elsevier, 1960	535	15,517	19
1960	Computing methods and the phase problem in X-ray crystal analysis: Glasgow Ed. Pepinsky,R., Robertson,J.M. & Speakman,J.C. -London: Pergamon, 1961	326	15,814	3
1960	Electron microscopy (Proceedings of European regional conference): Delft Vol.1: Ed. Houwink,A.L. & Spit,B.J. -Den Haag: Nederlandse Vereniging voor Electronenmicroscopie, 1961	606	15,1063	19,11
1960	Felspars: Copenhagen Symposium of the International Mineralogical Association [Published as Cursillos y conferencias VII] Ed. Laves,F. & Fuster,J.M. -Madrid: C.S.I.C., 1961	182		15,8
1960	Fermi surface: Cooperstown (N.Y.) Ed. Harrison,W.A. & Webb,M.B. -New York: Wiley, 1961	356		9,12
1960	Reactivity of solids (Proceedings of the fourth international symposium): Amsterdam Ed. Boer,J.H.de -Amsterdam: Elsevier, 1960	762		16,11
1960	Semiconductor physics: Prague Sponsor: Czechoslovak Academy of Sciences -New York: Academic Press, 1961	608		9
1960	Ferrity: fizičeskie i fiziko-khimičeskie svojstva. Doklady III vsesojuznogo soveščanija [Ferrites: physical and physico-chemical properties. Proceedings of the third national conference] Ed. Sirota,N.N., Belov,K.P. & others -Minsk: Akad.nauk.BSSR, 1960	652		8
1960/61	Properties of crystalline solids - recent progress in materials science: Philadelphia and Nature and origin of strength of materials: Philadelphia -Philadelphia (Penna.): ASTM, 1961	143		9
1961	Diffraction line profiles (The measurement and interpretation of): Bournemouth Ed. Rand,M.H. (for U.K. Atomic Energy Authority) -London: H.M.Stationery Office, 1962	78		5
1961	Direct observation of imperfection in crystals: St. Louis (Missouri) Ed. Newkirk,J.B. & Wernick,J.H. -New York: Interscience, 1962	617		11,19
1961	Electron microscopy and strength of crystals: Berkeley (Cal.) Ed. Thomas,G. & Washburn,J.W. -New York: Interscience, 1963	1002	17,1340	19,11

Year	Title, place, editor, publisher	No. of pages	Review ref.	Subj ref
1961	Magnetism and crystallography: Kyoto [Published as supplement to Vol.17 of J.Phys.Soc.Japan; ed. Miyake,S.] -Tokyo: Physical Society of Japan, 1962			
	B - I Magnetism	718	16,855	9
	B -II Electron and neutron diffraction	397	16,856	5
	B-III Neutron diffraction study of magnetic materials	71	16,236	9,
1961	Nucleation and crystallisation in glasses and melts: Toronto Ed. Reser,M.K., Smith,G. & Insley,H. - Columbus (Ohio) : American Ceramic Society, 1962	123		13,
1961	Scientific information in the fields of crystallography and solid state physics: Nishinomiya (Japan) Ed. Watanabé,T. & Takéuchi,W. -Crystallographic Society of Japan, 1962	131	16,155	T,
1961	Ultrastructure of protein fibres: Pittsburgh Ed. Borasky,R. -New York: Academic Press, 1963	185		17
1962	Advances in technique in electron metallography (Symposium on ...): New York -New York: ASTM, 1963	72		5,
1962	Crystal lattice defects: Tokyo and Kyoto [Published as supplement to Vol.18 of J.Phys.Soc.Japan] -Tokyo: Physical Society of Japan, 1963			11
	I Symposium	199		
	II Conference (1)	356		
	III Conference (2)	373		
1962	Feldspars (Advanced study institute): Oslo [Published as Vol.42, Part 2, of Norsk Geologisk Tidsskrift] Ed. Christie,O.H.J. -København: Fr.Bagges Kgl.Hofbogtrykkeri, 1962	606		15
1962	Fizika ščeločno-galoidnikh kristallov - Trudy 2-go soveščanija [Physics of alkali halide crystals - Proceedings of 2nd conference]: -Riga: Izdat.Univ., 1962	548		9
1962	Informational macromolecules: Rutgers Ed. Vogel,H.J., Bryson,V. & Lampen,J.O. -New York: Academic Press, 1963	542		17
1962	Magnetic and electric resonance and relaxation (Proceedings of the XIth Colloque Ampère): Eindhoven Ed. Smidt,J. -Amsterdam: North Holland, 1963	789		19
1962	Metallic solid solutions: Orsay (France) Ed. Friedel,J. & Guinier,A. -New York: Benjamin, 1963	[About 640]		11
1962	Physics of semi-conductors (Proceedings of international conference on ...): Exeter Ed. Stickland,A.C. -London: Institute of Physics, 1962 Chapman & Hall, 1963	909		9

Conferences - II

Year	Title, place, editor, publisher	No. of pages	Review ref.	Subject ref.
1962	Recovery and recrystallization of metals: New York Ed. Himmel,L. -New York: Interscience, 1963	389	17,1090	12
1962	Structure and properties of engineering materials: Raleigh (N.C.) [Published as Materials Science Research, Vol.1] Ed. Stadelmaier,H.H. & Austin,W.W. -New York: Plenum Press, 1963	300		11,9
1962	X-ray optics and X-ray microanalysis (Third international symposium): Stanford (Cal.) Ed. Pattee,H.H., Cosslett,V.E. & Engström,A. -New York: Academic Press, 1964	622		19
1963	Aspects of protein structure: Madras Ed. Ramachandran,G.N. -New York: Academic Press, 1963	382		17
1963	Crystallography and crystal perfection: Madras Ed. Ramachandran,G.N. -New York: Academic Press, 1963	374	17,1617	3,11,5
1963	Advanced methods of crystallography: Madras [winter school] Ed. Ramachandran,G.N. -London: Academic Press, 1964	280		3,11
1963	Single crystal films: Philco, Bluebell (Penna.) Ed. Francombe,M.H. & Sato,H. -Oxford: Pergamon, 1964	420		10

SERIAL PUBLICATIONS - III

Explanation

This term is used to cover multi-volume works where the separate volumes appear at intervals under a common title, or (sometimes) under the name of the original compiler of a work of reference which is now revised and rewritten by others (e.g. Dana's "System of Mineralogy"). Series of monographs in which each volume has its own title and its own author are not included, even when advertised by their publisher as a series; they are to be found under their authors in List I. Journals appearing in parts at regular intervals are altogether outside the scope of this Book List.

List III also includes the Conference Series mentioned in the Preface to List II.

List III is arranged alphabetically by either the title or the name of the original compiler (now effectively used as a title). The short title is often followed, (either after a dash or in brackets) by additional matter to help in identification. Details common to the whole series come next, and then details about separate volumes. Where the series is not primarily crystallographic, an attempt has been made to pick out relevant volumes.

For explanation of the remaining points, see the preface to List I.

In general, the purpose of this List is to separate out those multi-volume works which are more naturally thought of by crystallographers under their titles than under their editors, and which might otherwise be hard to find - even though many of the works concerned are both familiar and in widespread use. Cross references are given in all doubtful cases.

Title	Publisher, date, and other details	No. of pages	Review ref.
Advances in structure research by diffraction methods - see	Fortschritte der Strukturforschung mit Beugungsmethoden		
Advances in X-ray analysis [Denver Conferences]	Ed. Mueller,W.M.		
	-New York: Plenum Press		
	Vol.1: (6th Conference, 1957) 1960	494	14,442
	Vol.2: (7th Conference, 1958) 1960	359	14,442
	Vol.3: (8th Conference, 1959) 1960	376	14,442
	Vol.4: (9th Conference, 1960) 1961	568	15,625
	Vol.5: (10th Conference, 1961) 1962	564	16,156
	Vol.6: (11th Conference, 1962) 1963	480	17,791
	Vol.7: (12th Conference, 1963) 1964	700	
American Institute of Physics Handbook	Ed. Frederikse,H.P.R.		
	-New York: McGraw Hill, (2)1963		
	Section 9: Solid-state physics		
Barker Index of crystals	Porter,M.W. & Spiller,R.C.		
	-Cambridge (England): Heffer		
	Vol.1 Tetragonal, hexagonal, trigonal and orthorhombic systems, 1951		5,854
	Part 1, introduction and tables	350	
	Part 2, description	1068	
	Vol.2 Monoclinic system, 1956		10,486
	Part 1, introduction and tables	383	
	Part 2, crystal descriptions	1800	
	Part 3, crystal descriptions (continued)	1772	

Serial publications - III

Title	Publisher, date, and other details	No. of pages	Rev re
Clay and clay minerals (Proceedings of National Conferences on)	-Oxford: Pergamon		
	9th, 1961: Purdue Univ. ed. Swineford,A., 1962	614	16,
	10th, 1961: Austin (Texas) ed. Swineford,A., 1963	509	17,
	11th, 1962: Ottawa ed. Bradley,W.F., 1963	368	17,
Crystal data - see Donnay,J.D.H. & others, List I			
Crystal structures - see Wyckoff,R.W.G., List I			
Dana,J.W. & Dana,E.S. - The system of mineralogy	Rewritten by Palache,C., Berman,H. & Frondel,C. -New York: Wiley		
	Vol.1: Elements, sulfides, sulfosalts (7) 1954, repr.1958	834	
	Vol.2: Halides, nitrates, borates, carbonates, sulfates, phosphates, arsenates, tungstates, molybdates, etc., (7) 1951	1124	6,
	Vol.3: Silica minerals, (7) 1962	334	17,
Denver Conferences - see Advances in X-ray analysis			
Encyclopedic dictionary of physics	Ed. Thewlis,J. -Oxford: Pergamon 1961-63		
	Vol.1	800	15,
	Vol.2	880	16,
	Vol.3	894	16,
	Vol.4	836	16,
	Vol.5	782	16,
	Vol.6	883	
	Vol.7	866	
	Vol.8 (with indexes)	498	
Enrico Fermi, International School - see Italian Physical Society			
Ergebnisse der exakten Naturwissenschaften	Ed. Flügge,S. & Trendelenburg,F. -Berlin: Springer		
	Band 23, 1950		
	Band 26, 1952		
	Band 28, 1955		
	Band 33, 1961		
Fortschritte der Strukturforschung mit Beugungsmethoden (Advances in structure research by diffraction methods)	Ed. Brill,R. -Braunschweig: Vieweg		
	Vol.1, 1964	224	
Geochemische Verteilungsgesetze der Elemente - see Goldschmidt,V.M., List I			
Geological Society of America, Special papers - see under names of authors, List I			
Glassy state (The structure of glass) Proceedings of All-Union Conferences, Leningrad	(tr. from Russian by Uvarov,E.B. & others) -New York: Consultants Bureau		
	Vol.I: (2nd Conference, 1953) 1958	296	
	Vol.II: (3rd Conference, 1959) 1961	492	15
	Vol.III: - Catalysed crystallization of glass - ed. Porai-Koshits,E.A., 1964	216	

Serial publications - III

Title	Publisher, date, and other details	No. of pages	Review ref.
Growth of crystals (tr. from Russian, see Rost kristallov)			
	Ed. Shubnikov,A.V. & Sheftal,N.N.		
	-New York: Consultants Bureau		
	1st (1956) Conference, 1958	294	11,900
	Interim reports, 1959	178	14,331
	2nd (1959) Conference, 1962		
Handbuch der Physik			
	Ed. Flügge,S.		
	-Berlin: Springer		
	Bd.7, Teil 1: Kristallphysik I, 1955	687	9,620
	Bd.7, Teil 2: Kristallphysik II, 1958	273	12,481
	Bd.10: Structure of liquids, 1961	326	
	Bd.25, Teil 1: Kristalloptik, Beugung, 1961	616	
	Bd.30: Röntgenstrahlen, 1957	384	
	Bd.32: Strukturforschung, 1957	663	11,59
	Bd.33: Korpuskularoptik, 1956	702	
	Bd.34: Korpuskeln und Strahlung in Materie II, 1958	316	
Handbuch der Mineralogie - see Hintze			
Hintze,C. - Handbuch der Mineralogie			
	Bd.I.1: Elemente und Sulfide, 1904	1208	
	Bd.I.2: Oxyde und Haloide, 1915	1467	
	Bd.I.3 [with Linck,G.; in two halves]: Nitrate, Jodate, Karbonate, Selenite, Tellurite, Manganite, Plumbate; Sulfate, Chromate, Molybdate, Wolframate, Uranate, 1930	1891	
	Bd.I.4 [with Linck,G.: in two halves]: Borate, Aluminate und Ferrate, Phosphate, Arseniate, Antimoniate, Vanadate, Niobate, und Tantalate; Arsenite und Antimonite, organische Verbindungen, 1933	1454	
	Bd.II: Silicate und Titanate, 1897	1841	
	Ergänzungsband I [with Linck,G.]: Neue Mineralien, 1938	760	
	Ergänzungsband II [Chudoba,K.F.]: Teil I, 1954	480	
	Ergänzungsband II [Chudoba,K.F.]: Teil II, Teil III, 1960	478	
	-Leipzig: Von Veit (before 1930)		
	-Berlin: de Gruyter (1930 and after)		
International encyclopedia of physical chemistry and chemical physics			
	Ed. Guggenheim,E.A., Mayer,J.E. & Tompkins,F.C.		
	Topic 11: The ideal crystalline state		
	Ed. Blackman,M.		
	-London: Pergamon, 1963	1	
International tables for X-ray crystallography			
	Ed. Lonsdale,K. & others		
	-Birmingham (England): Kynoch Press		
	Vol.1: Symmetry groups, ed. Henry,N.F.M. & Lonsdale,K., 1952	558	7,304
	Vol.2: Mathematical tables, ed. Kasper,J.S. & Lonsdale,K., 1959	444	
	Vol.3: Physical and chemical tables, ed. MacGillavry,C.H. & Rieck,G.D., 1962	368	16,234
Internationale Tabellen zur Bestimmung von Kristallstrukturen			
	Ed. Hermann,C. (with Bragg,W.H. & Laue,M.von)		
	-Berlin: Borntraeger, 1935		
	Bd.I: Gruppentheoretische Tafeln	451	
	Bd.II: Mathematische und physikalische Tafeln	241	
Italian Physical Society - Proceedings of international school of physics, 'Enrico Fermi'			
	-New York: Academic Press		
	Course 18: Radiation damage in solids (Ispra 1960) Ed. Billington,D.S., 1962	942	
	Course 22: Semi-conductors (Varenna 1961) Ed. Smith,R.A.	540	

Serial publications - III

Title	Publisher, date, and other details	No. of pages	Revi ref
Kristalličeskie struktury (perev. s angl. - novye issledovanija)			
	[Crystal structure - tr. from English - new investigations]		
	Ed. Bokij,G.B.		
	-Moskva: MIR		
	Sb.3: (3rd collection), 1951	168	
	Sb.4: (4th collection), 1951	312	
Kristallografija [Crystallography]			
	Ed. Shafranovskij,I.I. & Mikheev,V.I.		
	Vypusk 1: Trudy fedorovskoj naučnoj sessii 1949 -Moskva: Metallurgija, 1951.	251	
	Vypusk 2: Trudy fedorovskoj naučnoj sessii 1951 -Ugletekhizdat, 1952	264	
	Vypusk 3: Posvjaščen stoletiju so dnja roždenija akademika E.S.Fedorova -Leningrad: Izdat.Univ., 1955	256	
	Vypusk 4: Trudy fedorovskoj naučnoj sessii 1953 -Leningrad: Izdat.Univ., 1955	232	
	Vypusk 5: - Moskva : Metallurgija, 1956	286	
	(Parts 1, 2, 4: Proceedings of the Fedorov Scientific Session; Part 3, in celebration of the centenary of the birth of Academician E.S.Fedorov)		
Kristallografija [Crystallography]			
	Sbornik statej posvjaščennykh pamjati Prof.Mikheeva,V.I. (Collection of papers in memory of Prof.V.I.Mikheev)		
	Ed. Levenberg,N.V.		
	-Moskva: NEDRA, 1961.	190	
Landolt-Börnstein - Zahlenwerte und Funktionen (Sechste Auflage der "Physikalisch-chemischen Tabellen")			
	-Berlin: Springer		
	Bd.1, Teil 4: Kristalle, 1954	1007	2,
	Bd.2, Teil 3: Schmelzgleichgewichte und Grenzflächenerscheinungen, 1957	535	
	Bd.2, Teil 8: Optische Konstanten, 1962	901	
	Bd.4, Teil 3: Elektrotechnik, Lichttechnik, Röntgentechnik, 1957	1076	
Materials Science Research, Vol.1 - see Conference 1962 "Structure and properties of engineering materials", List II			
Partington,J.R. - An advanced treatise on physical chemistry			
	-London: Longmans		
	Vol.III: The properties of solids (including crystallography), 1952	639	
	Vol.V: Molecular spectra and structure, dielectrics and dipole moments, 1953	688	
Proceedings of the Institute of Crystallography, Moscow (in Russian) - see Trudy instituta kristallografii			
Progress in biophysics			
	-London: Pergamon		
	Vol.5: (ed. Butler,J.A.V. & Randall,J.T.), 1955	231	
	Vol.7: (ed. Butler,J.A.V. & Katz,B.), 1957	362	
Progress in crystal physics			
	Ed. Krishnan,R.S.		
	-New York: Interscience		
	Vol.1, 1958	198	
Progress in low temperature physics			
	Ed. Gorter,G.J.		
	-Amsterdam: North Holland		
	Vol.4, 1964	566	

Serial publications - III

Title	Publisher, date, and other details	No. of pages	Review ref.
Progress in materials science (Continuation of Progress in metal physics)	Ed. Chalmers,B.		
	-Oxford: Pergamon		
	Vol.8		
	Vol.9, 1961	386	
	Vol.11, 1963		
Progress in metal physics	Ed. Chalmers,B. & King,R.		
	-London: Butterworths (Vols.1-2)		
	-Oxford: Pergamon (Vols.3-7)		
	Vol.1, 1949	401	
	Vol.2, 1950	211	
	Vol.3, 1952	334	
	Vol.4, 1953	403	
	Vol.6, 1956	354	
Progress in stereochemistry	Ed. de la Mare,P.B.D. & Klyne,W.		
	-London: Butterworths		
	Vol.2, 1958	323	
Reports on progress in physics	Ed. Stickland,A.C.		
	-London: Institute of Physics		
	Vol.18, 1953	407	7,143
	Vol.24, 1961	424	15,624
	Vol.25, 1962	529	17,72
	Vol.26, 1963		17,791
Rost kristallov - Doklady na soveščanii porosty kristallov (Proceedings of conference(s) on growth of crystals - for English translation of series see Growth of crystals)	Ed. Shubnikov,A.V. & Sheftal,N.N.		
	-Moskva: NAUKA		
	Vol.1, 1956		
	Vol.2, 1959		
	Vol.3, 1961		
	Vol.4, 1964		
Solid state physics	Ed. Seitz,F. & Turnbull,D.		
	-New York: Academic Press		
	Vols.1-12, 1955-1961	[Each about 500]	11,451
	Vols.13-14, 1962-1963	[Each about 500]	
Solid state physics supplements - see under authors in List I			
Solid state physics, reprints from - see under authors in List I			
Soviet research in crystallography (in English translation)	-New York: Consultants Bureau		
	Chemistry collection No.5, Vol.2, 1958	230	12,80
Structure of glass - see Glassy state			
Structure reports (International Union of Crystallography)	Ed. Wilson,A.J.C. (Vols.8-18) & Pearson,W.B. (Vols.19-21)		
	-Utrecht: Oosthoek		
	Vol.8: (1940-1941), 1956	384	10,388
	Vol.9: (1942-1944), 1955	448	9,840
	Vol.10: (1945-1946), 1953	325	8,70
	Vol.11: (1947-1948), 1951	779	5,299
	Vol.12: (1949), 1952	478	6,671
	Vol.13: (1950), 1954	643	8,444
	Vol.14: (1940-1950 supplement, and cumulative indexes), 1959	214	

(cont.)

Title	Publisher, date, and other details	No. of pages	Rev[illegible] ref[illegible]
	Vol.15: (1951), 1957	588	11,[illegible]
	Vol.16: (1952), 1959	651	
	Vol.17: (1953), 1963	863	17,[illegible]
	Vol.18: (1954), 1961	845	
	Vol.19: (1955), 1963	690	
	Vol.20: (1956), 1963	728	
	Vol.21: (1957), 1964	706	
Strukturbericht (Zeitschrift für Kristallographie Ergänzungsband)			
	-Leipzig: Akademische Verlagsg.;		
	-Ann Arbor (Michigan): Edwards, repr. 1943		
	-New York: Johnson Reprint Corporation, repr. 1964		
	Bd.1: (1913-1928) Ed. Ewald,P.P. & Hermann,C., 1931	815	
	Bd.2: (1928-1932) Ed. Hermann,C. & others, 1937	963	
	Bd.3: (1933-1935) Ed.Gottfried,C., 1937	899	
	Bd.4: (1936) Ed. Gottfried,C., 1938	346	
	Bd.5: (1937) Ed. Gottfried,C., 1940	184	
	Bd.6: (1939) Ed. Herrmann,K., 1941	285	
	Bd.7: (1939) Ed. Herrmann,K., 1943	305	
Trudy instituta Kristallografija - periodičeskie sborniki			
	[Proceedings of the Institute of Crystallography, periodical collections]		
	-Moskva: NAUKA		
	Vypusk 1, 1948 - Vypusk 12, 1956		
	(Parts 1-12, 1948-1956)		
World directory of crystallographers - see Smits, D.W. (ed.), List I			

SUBJECT CLASSIFICATION - IV

Books are here arranged alphabetically by author (or editor) under the following main subject headings.

1. General crystallography
2. Geometry and symmetry of crystals and periodic structures
3. Structure analysis (including Fourier and "direct" methods)
4. Diffraction theory
5. Diffraction techniques and applications (including electron and neutron diffraction)
6. Mathematical treatments
7. Crystal chemistry
8. Chemical and physical properties of particular materials
9. Crystal physics
10. Surfaces and thin films
11. Imperfections (including order-disorder relations, dislocations, and radiation damage)
12. Metals and metallic textures
13. Morphology and growth of crystals
14. Crystal optics
15. Mineralogy
16. Changes of state, phase transitions and diffusion processes
17. Large molecules (including crystalline polymers, proteins, and materials of biological importance)
18. Non-crystalline and partly crystalline materials (including liquids and glasses)
19. Various techniques (including electron and X-ray microscopy, spectroscopy and resonance methods)
20. Historical
21. Miscellaneous (including symmetry in art and nature, and philosophical implications of crystallography)

T. Tables, atlases, literature surveys, and data compilations

Further amplification of these main divisions, with cross-references and notes about overlaps, is given at the heads of the separate class lists.

For many of the main divisions, a sub-classification has been added, by means of a letter preceding the author's name, whose meaning is explained immediately below the main heading. Absence of a letter means either that the book covers a wide range of topics under the main heading, or that no sub-heading describes it well, or that its correct sub-classification has not been decided.

Few of the boundaries can be sharply drawn; many books have been listed under two or even three main headings.

Entries are generally in very abbreviated form, intended to be used in conjunction with Lists I - III, where details are given. For List I entries only the author's (or editor's) name is given; for List II entries (Conferences) the date is added. List III entries follow at the foot of each class list.

For List IV.T, Tables and data compilations, the wide range of subject matter and aims made a less abbreviated list desirable. Here, titles or shortened titles are given (for Russian works, translations or paraphrases of the title, in square brackets).

List IV. Preface continued

Two other lists are added, of classes of books picked out from the Main List. List IV.X gives the old books, up to 1930; List IV.Y gives the "popular" books, with an appeal to a wider public. The latter comprise some, but by no means all, of the books of teaching level E or EU, some of the books so classed being suitable only for the serious beginner and not for general reading. For neither class was there any comprehensive search made or attempt at complete coverage; the selection is probably arbitrary and incomplete, but may provide a starting point for those interested.

1. General crystallography (including general textbooks, and collections of papers)

Anseles
Barker, T.V.
Barraud
Belaiew
Belov, N.V.
Bijvoet, Kolkmeijer & MacGillavry
Bokij & Porai-Koshits
Boldyrev
Bouman
Bragg, W.H.
Bragg, W.H. & Bragg, W.L.
Bragg, W.L.
Brasseur
Broglie, M.de
Bruhns & Ramdohr
Buerger
Bunn
Clark
Davey
Dolivo-Dobrovol'skii
Fedorov, E.S.
Flint
Flint & Anvaer
Friedel, G.
Garratt
Garrido
Garrido & Orland
Gay
Glocker
Guinier
Guinier & Dexter
Hauy
Hirst
Holden & Singer
James
Jong, de
Jong, de & Bouman
Kitaigorodskii
Kleber
Kohlhaas & Otto
Kožina, Stroganov & Tolkačev
Lonsdale
McLachlan
Mauguin
Neff
Niggli
Padurov
Pfeiffer
Popov & Shafranovskii
Porai-Koshits
Raaz & Kohler
Raaz & Tertsch
Samoilov
Shaskolskaya
Shubnikov
Shubnikov, Flint & Bokij
Thibaud
Trapeznikov
Trey & Legat
Trillat
Tunnell & Murdoch
Tutton
Winkler
Wohlrabe
Wulff
Ždanov

See also serial publications (List III):

Advances in X-ray analysis
Fortschritte der Strukturforschung mit Beugungsmethoden
Kristallografija

List IV

2. Geometry and symmetry of crystals and periodic structures (including point groups and space groups treated geometrically, and colour groups, but not general group theory - for which see 6)

a : macroscopic symmetry and geometry
b : periodic structures
T : tables or compilations of data

	Aňseles
	Barker, T.V.
	Barker Index (List III)
b	Belov, N.V.
a	Bogomolov
	Boldyrev & others
a	Bonštedt
b	Bravais
b	Buerger
b	Burckhardt
	Cundy & Rollett
b	Delone, Padurov & Aleksandrov
a	Donnay
b	Faddeev
	Fedorov, E.S.
a	Fischer
	Flint
	Friedel, G.
	Gadolin
	Garrido
T	Goldschmidt, V.
b	Hilton
	Jaeger
b	Koster
a	Koster & others
	Levinson-Lessing
T	Lonsdale
	Niggli
a	Phillips
b	Rinne & Schiebold
	Romanov
b	Schieltz
	Schoenflies
a	Schroeder
a	Shubnikov
T	Shubnikov
	Shubnikov, Belov & others
	Sohncke
a	Sommerfeldt
a	Terpstra
b	Terpstra
a	Terpstra & Codd
a	Tertsch
	Tutton
	Vernadskii
T	Wolf & Wolff
	Wood
	Wulff
a	Wulff & Shubnikov
b	Wyckoff

3. Structure analysis

a : Fourier analysis of crystal structures
b : "Direct" methods

	Barber
a	Booth
a	Brasseur
b	Hauptman & Karle
b	Kitaigorodskii
a	Lipson & Cochran
a	Lipson & Taylor
a	Nowacki
a	Nyburg
	Pepinsky (1952)
	Pepinsky & others (1960)
	Pines, B.Ja.
	Ramachandran (1963)
a	Robertson
a	Taylor & Lipson
	Vainštein
b	Woolfson
a	Wrinch

List IV

4. Diffraction theory (for experimental work see 5)

a : general
b : emphasis on imperfect materials

a Ewald
b Guinier
b Hosemann & Bagchi
a James
a Laue, von
b Pirenne
b Vainštein
b Wilson
b Wooster
a Zachariasen

Also serial publication (List III):
Handbuch der Physik, 25.

5. Diffraction techniques and applications (including electron diffraction and neutron diffraction, but excluding books dealing only or mainly with results, for which see later headings)

a : X-ray diffraction, general
b : single-crystal techniques
c : powder techniques; polycrystalline material
d : application to metals
e : application to imperfect and non-crystalline materials
f : electron diffraction
g : neutron diffraction
T : tables or compilations of data

d Ageev
a Amoros
f ASTM (1962)
c Azaroff & Buerger
g Bacon
d Bagarjackij (ed.)
d Barrett
f Bauer
f Beeching
T Berry & Thompson
T Boldyrev & others
b Buerger
T Cauchois & Hulubei
a Cullity
a Dauvillier
T Dettmar & Kirchner
c D'Eye & Wait
e Dubinin (ed.)
a Frank-Kameneckii & others
b Furnas
d,T Gorelik & others
a,e Guinier
a Guinier & Dexter
e Guinier & Fournet
a Henry, Lipson & Wooster
a Huerta
g Hughes
a Khejker & Zevin
e Kitaigorodskii
c Klug & Alexander
T König
a Kurdumov (ed.)
a Mark
f Mark & Wierl
T Mirkin
c Parrish (ed.)
T Parrish & others
c Peiser & others (eds.)
f,g Physical Society of Japan (19
a Pines, B.Ja.
f Pinsker
a Ramachandran (1963)
c Rand (1961)
e Randall
T Rose
b Schiebold

List IV

5. continued

b Schieltz
T Shell
c Straumanis & Ievins
d Taylor, A.
T Taylor & Kagle
T Tolkačev
d Umanskij, Ja.S.
a Umanskij, Ja.S. & others
a Umanskij, M.M.
f Vainštein
e Wilson
b Wood
d Ždanov & Umanskij

See also serial publications (List III):
Handbuch der Physik, 30,33,34;
and Tables in List IV.T

6. Mathematical treatments (including group theory and other mathematical analyses not coming under 2,3 or 4).

a : Group theory
b : atomic form factors
c : Fourier theory, not applied to crystals

c Barber
Bethe
a Bhagavantam & Venkataraydu
a Burckhardt
a Faddeev
b Hartree
b Herman & Skillman
a Hilton
c Jennison
a Kovalev
a Ljubarskij
a Lomont
a McWeeny
a Mariot
a Matossi
a Pines, D.
a Speiser

7. Crystal chemistry. (For chemical properties of particular materials, see 8,17)

a : general
b : inorganic structures
c : organic structures
d : metallic structures
e : molecular geometry and geometry of continuously-linked structures
f : chemical bonds
g : ligand-field theory
h : solid-state chemistry
i : magnetochemistry
j : non-stoichiometry
k : geochemistry
T : tables or compilations of data

Addison
a Amoros
a Arkel, van
a Bacon
a Belov, N.V.
g Bersuker & Ablov
i Bhatnagur & Mathur
a Bokij
b Bragg, W.L.
f Brand & Speakman
a Brandenberger
a Brandenberger & Epprecht
c Brasseur
e Broglie, L.de (1953)

7. continued

a	Bunn	b	Megaw
f	Coulson	a	Niggli
f	Enz	a	Nowacki
a	Evans	c	Nyburg
h	Fermi	g	Orgel
c	Fox & others (ed.)	b	Ormont
h	Garner	f	Pauling
b,k	Goldschmidt, V.M.	a	Raaz & Kohler
i	Goodenough	h	Rees
g	Griffith	c	Robertson
T	Groth		
a	Hassell	e	Sanderson
h	Hedvall	b,T	Schubert
f	Hein	f	Seel
a	Hevesy	f	Slater
a	Hiller	f	Speakman
a	Hocart & Kern	T	Strunz
d	Hume-Rothery	T	Sutton (ed.)
b	Hückel	T	Timmermans
T	IUPAC	a	Tutton
f	Ketelaar	b,e	Wells
c	Kitaigorodskii	a	Wheatley
a	Kleber	T	Winchell
h	Kröger	e	Wolkenstein
j	Mandelcorn (ed.)	a,T	Wyckoff

See also serial publications (List III):

Progress in stereochemistry
Structure Reports
Strukturbericht

8. Chemical and physical properties of particular materials

a : diamond
b : silicates
c : cement compounds
d : oxide magnetic materials

b	Belov, N.V.	b	Laves & Fuster (1960)
b	Bragg, W.L.		Lipscomb
c	Cement and Concrete Association, (1952)		Makarov
a	Champion	a	Rodewald
b	Christie (1962)	a	Shafranovskii
b	Eitel	d	Sirota & others (1960)
a	Fersman	d	Standley
a	Fersman & Goldschmidt	c	Taylor, H.F.W. (ed.)
b	Grofcsik & Tamas	a	Tolansky
	Gschneider		Ubbelohde & Lewis
	Hagan		Wrinch
c	Heller & Taylor		

List IV

9. Crystal physics

a : general, mainly macroscopic
b : general "solid-state"
c : electronic properties
d : lattice dynamics
e : dielectrics, piezoelectricity and ferroelectricity
f : magnetic properties (for resonance see also 19)
g : luminescence and optical absorption (for refraction and classical optics see 14; for spectroscopy see 19)
h : acoustic oscillations (including ultrasonics)
i : elasticity, plasticity and deformations (see also 11)
j : materials science

	Amoros
i,j	ASTM (1960/1)
b	Azaroff
c	Azaroff & Brophy
f	Belov, K.P.
h	Bergmann
d	Bhagavantam & Venkataraydu
f	Bhatnagar & Mathur
e	Birks (ed.)
f	Birss
i	Boas
d	Born
d	Born & Huang
d	Brand & Speakman
d	Brillouin
d	Brillouin & Parodi
d	Broglie, L.de (1953)
a	Cady
c	Callaway
c	Champion
i	Cottrell
g	Curie
c	Czechoslovak Academy (1960)
d	Davydov
b	Dekker
d	Fermi
i	Fisher & others (1956)
j	[Fizika ščeločno-galoidnikh kristallov] (1962)
j	Fox & others (ed.)
j	Fréchette (1958)
j	Garner
i	Gercriken & others
f	Goodenough
a	Groth
c	Harrison & Webb (1960)
h	Herzog
j	Hippel, von
i	Houwink
j	Hutchison & Baird
i	Huntingdon
a,c	Ioffe
i	Jaswon
e	Jaynes
e	Jona & Shirane
c	Jones, H.
c	Justi
e	Känzig
i	Klassen-Neklyudova (ed.)
	Kleber
i	Kochendörfer
i	Kochendörfer & Seeger
b	Koerber
a	Liebisch
i	Likhtman & others
	Ljubarskij
	McClure
d	Maradudin & others
e	Megaw
g	Moss
c	Mott & Gurney
c	Mott & Jones
a	Nye
i	Paul & Warschauer (eds.)
b	Peierls
c	Pekar
i	Pen'kovskii
i	Physical Society, London (1947)

9. continued

f	Physical Society of Japan (1961)	c	Smith, R.A. (ed.)
d	Pines, D.	j	Stadelmaier & Austin (1962)
c	Post	f	Standley
g	Przibram	c	Stasiw
f	Rado & Suhl	c	Stickland (1962)
c	Raimes	c	Stumpf
b	Rees	a	Swalin
b	Sachs	i	Tertsch
e	Sachse	j	Van Vlack
i	Savitsky	a	Voigt
i	Schmidt & Boas	h	Waller
c	Shockley	i	Westbrook (ed.)
e	Shubnikov	j	Wolkenstein
e	Shubnikov & others	a	Wooster
f	Sirota & others (1960)	c	Ziman
c,d	Smith, R.A.	b	Ždanov

See also serial publications (List III):

Ergebnisse der exakten Naturwissenschaften
Handbuch der Physik
Progress in crystal physics
Progress in low temperature physics
Reports on progress in physics
Solid state physics

10. Surfaces and thin films

Chalmers	Kuznecov
Deryagin (ed.)	Neugebauer & others (1959)
Gomer & Smith (1952)	Tolansky

11. Imperfections

a : order-disorder relations (see also 4 and 16)
b : dislocations
c : radiation damage

(See also 9 for point defects, 13 for coarser defects in general, 12 for metallic textures, 18 for non-crystalline materials, and 5 for experimental investigations)

c	Billington (ed.)	a	Elcock
c	Billington & Crawford (eds.)	b	Fisher & others (1956)
b	Boer, de (1960)	b	Friedel, J.
	Bueren, van		Friedel & Guinier (1962)
c	Cauchois (1956)	a	Green & Hurst
b	Cottrell		Guinier
b	Dekeyser & Amelinckx		Guinier & Fournet
c	Dienes & Vineyard (eds.)		Hosemann & Bagchi
	Doremus & others (1958)	b	Houwink

11. continued

b	Houwink & Spit (1960)		Rassweiler & Grube (1958)
	Klassen-Neklyudova (ed.)	b	Read
	Krivoglaz & Smirnov		Shockley & others (1950)
	Kröger		Stadelmaier & Austin (1962)
a	Muto, Takagi & Guttman		Stokes
	Newkirk & Wernick (1962)		Thomas & Washburn (1961)
	Pen'kovskii	b	Vasil'ev
	Physical Society, London (1947)	b	Verma
	Physical Society, London (1954)		Wilson
	Physical Society of Japan (1962)		Wooster
	Ramachandran (1963)		

See also serial publications (List III):

Progress in metal physics
Progress in materials science

12. Metals

a : general
b : structures
c : theoretical treatments (see also 9)
d : properties (see also 9)
e : textures and imperfections (see also 11)
T : tables or compilations of data

a	Ageev		Krivoglaz & Smirnov
e	Arkharov	e	Kurdjumov
	ASTM (1962)	d	Likhtman & others
a	Bagarjackij (ed.)	e	Mott
a	Barrett	c	Mott & Jones
e	Belaiew	c	Muto, Takagi & Guttman
	Belov, N.V.	T	Pearson
d	Boas	c	Raimes
c	Cottrell	e	Rassweiler & Grubb (1958)
e	Fast & others (1956)	d	Savitsky
e	Friedel & Guinier (1962)	d	Schmidt & Boas
e	Gercriken & others	a	Smallman
a	Gorelik & others	T	Smithells
	Gschneider	a	Taylor, A.
e	Guinier	T	Taylor, A. & Kagle
e	Hall	a	Umanskij, Ja.S.
d	Halla	e	Vasil'ev
c	Harrison & Webb (1960)	e	Wassermann
e	Himmel (1962)	e	Wassermann & Grewen
b,c	Hume-Rothery		Westbrook (ed.)
b	Hume-Rothery & Raynor	c	Ziman
d	Kochendörfer	a	Ždanov & Umanskij, Ja.S.
d	Kochendörfer & Seeger		

See also serial publications (List III):

Progress in metal physics
Progress in materials science

List IV

13. Morphology and growth of crystals (including twinning in general, but excluding books mainly concerned with metals (12) or dislocations (11)).

Arkharov
Bentley & Humphreys
Buckley
Cinzerling
Deicha
Dekeyser & Amelinckx
Doremus & others (1958)
Faraday Society (1949)
Fisher & others (1956)
Gilman
Goldschmidt, V.
Holden & Singer
Honigmann
Kapustin
Klassen-Neklyudova
Kurdjumov (ed.)
Kuznecov
Lawson & Neilsen
Lemmlejn
Mandelkern
Mullin
Nakaya
Reser & others (1961)
Saratovkin
Shafranovskii
Shubnikov
Sirota (ed.)
Smakula
Van Hook
Verma
Wilke
Wulff

See also serial publications (List III):

Ergebnisse der exakten Naturwissenschaften
Rost kristallov (Growth of crystals)

14. Crystal optics (see also 9 for luminescence and absorption; 15 for optical mineralogy and 19 for spectroscopy)

Anseles & Burakova
Beljankin & Petrov
Bloss
Bokii
Buchwald
Burri
Cagnet, Françon & Thrierr
Fedorov, F.I.
Fletcher
Hartshorne & Stuart
Johannsen
Kordes
Kordes
Korenman
Moss
Pockels
Rinne & Berek
Sarančina
Shubnikov
Skobel'cin
Stokes
Vardanjanc
Wahlstrom
Winchell
Wood
Zavarickij (ed.)

15. Mineralogy

T : Tables or compilations of data

Belov, N.V.
Berry & Mason
T Berry & Thompson
Betekhtin
Bragg, W.L.
Brauns & Chudoba
T Brown, G.
Christie (1962)

List IV

15. continued

Correns
Dana
Deer, Howie & Zussman
Eitel
Eskola
Fersman
Fersman & Goldschmidt
Frank-Kameneckii & others
Goldschmidt, V.M.
Grigor'ev
Grim
Grofcsik & Tamas
T Heller & Taylor
T Hey
Hintze
Ito
Ketsmets
Kraus, Hunt & Ramsdell
Lapadu-Hargues
Laves & Fuster (1960)
Ložnikova & Jakovleva
Machatschki
Mikheev
Niggli
Nowacki
O'Daniel (1951)
Oelsner
T Philipsborn, von
Picon & Flahaut
Przibram
Ramdohr
Rodewald
Shafranovskii
Sobolev
Société Française de minéralogie
T Strunz
Taylor, H.F.W. (ed.)
Tolansky
Vardanjanc
Wade & Mattox
Wahlstrom
Walton
Webster
Werner
Winchell
T Winchell

See also serials publication (List III):
Clays and clay minerals

16. Changes of state, phase transformations, and diffusion (see also 9 for ferroelectricity and 11 for order-disorder relations)

Belov, K.P.
Boer, de (1960)
Bowden & Yoffe
Chalmers
Commissariat à l'énergie atomique (1958)
Eitel
Fast & others (1956)
Green & Hurst
Hall
Hauffe
Hedvall
Kapustin
Muto, Takagi & Guttman
[Reactivity of solids] (1952)
Sirota (ed.)
Smoluchowski & others (1948)
Temperley

17. Large molecules (including crystalline polymers, proteins, and materials of biological importance)

Borasky (1961)
Brown, N.R. (1954)
Engström
Engström & Finean
Geil
Mandelkern
Melville
Meyer & Mark

17. continued

Perutz

Ramachandran (1963)

Segal, Dornberger-Schiff & Kalaidjiev

Spiegel-Adolf & Henny

Springall

Vainštein

Vogel, Bryson & Lampen (1962)

Wolf

Wrinch

See also serial publication (List III):
Progress in Biophysics

18. Non-crystalline and partly crystalline materials

a : liquids
b : liquid crystals
c : glasses and non-crystalline solids

a Barker

a Bernal

a Danilov

c Dubinin (ed.)

b Ewald (ed.)

c Fréchette (1958)

b Gray

c Jones, G.O.

c Porai-Koshits (ed.)

c Randall

c Reser & others (1961)

See also serial publications (List III):
Glassy state
Handbuch der Physik 10

19. Various techniques

a : electron microscopy
b : X-ray microscopy
c : X-ray spectroscopy
d : optical, infrared and ultraviolet spectroscopy
e : resonance methods

e Andrew

ASTM (1962)

d Bhagavantam & Venkataraydu

Braude & Nachod

c Cauchois

c Cauchois & Hulubei

b Cosslett, Engström & Pattee (1956)

b Cosslett & Nixon

e Das & Hahn

Engström, Cosslett & Pattee (1959)

a Houwink & Spit (1960)

d McClure

a Möllestedt & others (1958)

Newkirk & Wernick (1962)

Pattee, Cosslett & Engström (

e Smidt (1962)

a Thomas

a Thomas & Washburn (1961)

Wheatley

List IV

20. Historical (including biographies, papers of historical interest, and collected papers)

Bragg, W.L.
Bravais
Ewald
Fletcher
Frisch, Paneth, Laves & Rosbaud (eds.) [Laue celebration]
Gadolin
Glasser
Hauy
Laue, von
Morse
Shafranovskii
Steno
Werner
Wulff

See also serial publications (List III):

Kristallografija, Pt.3 (1955) Fedorov centenary
(1961) Mikheev commemoration

21. Miscellaneous (including books on symmetry in art and nature, and philosophical implications of crystallography)

Bentley & Humphreys
Cundy & Rollett
Escher
Fersman
Jaeger
Niggli
Shubnikov
Terpstra
Thompson, D.W.
Thompson & Bonner
Tolansky
Waller
Watanabé & Takéuchi (1962)
Weyl

T. Tables, atlases, literature surveys and data compilations

Berry & Thompson	X-ray powder data for ore minerals
Boldyrev & others	[Handbook of crystals, Vol.1, Section 1]
Boldyrev, Mikheev & Dubinina	[Tables of spacings for Fe, Cu, Mo]
Brown, G.	X-ray identification and crystal structures of clay minerals
Buerger	Numerical structure factor tables
Cauchois & Hulubei	Longueurs d'onde des émissions X et des discontinuités d'absorptions X
Dana & Dana	The system of mineralogy
Dettmar & Kirchner	Tabellen zur Auswertung der Röntgendiagramme von Pulvern
Dolivo-Dobrovol'skii & others	[Handbook of crystals, Vol.1, Section 2]
Donnay & Donnay	Crystal data
Donnay, Nowacki & Donnay	Crystal data
Fadeev	[Tables of fundamental unitary representations of the Fedorov groups]
Fedorov, E.S.	Das Kristallreich
Goldschmidt, V.	Kristallographische Winkeltabellen
Goldschmidt, V.	Atlas der Kristallformen
Gorelik & others	[Tables and diagrams for analysis of metals by X-ray and electron diffraction]

T. continued

Groth	Chemische Kristallographie
Gschneider	Rare earth alloys
Heller & Taylor	Crystallographic data for the calcium silicates
Hey	An index of mineral species
Hintze	Handbuch der Mineralogie
Hodgman	Handbook of chemistry and physics [Rubber handbook]
IUPAC	Manual of physico-chemical symbols and terminology
IUPAC	Rules for notation for organic compounds
Kitaigorodskii	[Tables for X-ray structure analysis]
Knaggs, Karlik & Elam	Tables of cubic crystal structure
Kordes	Optische Daten zur Bestimmung anorganischer Substanzen
König	Anorganische Pigmente und Röntgenstrahlen
Lonsdale	Simplified structure factor and electron density formul
Ložnikova & Jakovleva	[Identification of minerals containing rare-earth eleme
Mirkin	Handbook of X-ray structure analysis of polycrystalline materials
Morse	Bibliography of crystal structure (1912-1927)
Parrish & others	Data for X-ray analysis
Partington	Advanced treatise on physical chemistry
Pearson	Handbook of lattice spacings of metals and alloys
Philipsborn, von	Tafeln zum Bestimmen der Minerale nach äusseren Kennzei
Rose	Tables...des diagrammes de rayons X et...monochromateur lame courbe
Rose (ed.)	Index of manufacturers
Sagel	Tabellen zur Röntgenstrukturanalyse
Sagel	Tabellen zur Röntgen-emissions und Absorptionsanalyse
Schubert	Kristallstrukturen zweikomponentiger Phasen
Shell	X-ray diffraction patterns of lead compounds
Shubnikov	[Atlas of crystallographic symmetry groups]
Smithells	Metals reference book
Smits (ed.)	World directory of crystallographers
Strunz	Mineralogische Tabellen
Sutton (ed.)	Tables of interatomic distances
Taylor & Kagle	Crystallographic data on metals and alloy structures
Timmermans	Physico-chemical constants of pure organic compounds
Tolkačev	[Tables of interplanar spacings]
Watanabe & Takeuchi	Scientific information in the fields of crystallography solid-state physics
Winchell	Microscopical characters of artificial inorganic solid substances
Winchell	Optical properties of organic compounds
Winchell	Elements of optical mineralogy
Wolf & Wolff	Symmetrie
Wyckoff	Crystal structures

See also serial publications (List III):
American Institute of Physics Handbook
Barker Index

List IV

T. continued

Encyclopedic Dictionary of Physics
International Tables for X-ray Crystallography
Internationale Tabellen zur Bestimmung von Kristallstrukturen
Landolt-Bornstein
Structure Reports
Strukturbericht

X. Older books: books first published in or before 1930, and not constantly reprinted - including books reprinted after a long lapse of time for their historical importance, with date of reprint in brackets

Barker, T.V., 1922, 1930, 1930
Belaiew, 1922
Born, 1923
Bragg, W.H., 1928
Bragg, W.H. & Bragg, W.L., 1915
Bravais, 1850 (1949)
Broglie, M.de, 1922
Dauvillier, 1924
Ewald, 1923
Fedorov, 1886, 1891, 1897, 1920
Fersman & Goldschmidt, 1911
Fletcher, 1892
Friedel, G., 1904, 1911 (1964)
Gadolin, (1954)
Goldschmidt, V., 1897, 1913-26
Goldschmidt, V.M., 1923-27
Groth, 1905, 1906-19
Haüy, 1784 (1962)
Hilton, 1903
[Ioffe=] Joffe, 1928, 1929
Jaeger, 1917
Johanssen, 1914 [still in print]
Laue, von, 1923
Levinson-Lessing & Beljankin, 1923
Liebisch, 1891
Mark, 1926
Mauguin, 1924
Meyer & Mark, 1930
Morse, 1928
Niggli, 1928
Pockels, 1906
Rinne & Schiebold, 1929
Schoenflies, 1891, 1923
Sohncke, 1876, 1879
Sommerfeldt, 1911
Steno, 1669 (1957)
Trillat, 1930
Tutton, 1911, 1924, 1926 (1964)
Vernadskii, 1904
Voigt, 1910
Werner, 1774 (1962)
Wulff, 1917, 1923
Wulff & Shubnikov, 1924
Wyckoff, 1922

Y. Popular books: books intended to interest a wider public. These may be identified in List I by looking for books marked as teaching level E or EU. Some other E or EU books not listed here may have been inadvertently omitted, but some are only intended for readers beginning a serious study of crystallography.

Belaiew
Bunn
Dana & Hurlbut
Garratt
Holden & Singer
Kitaigorodskii
Melville
Mott
Niggli
Pfeiffer
Raaz & Kohler
Shaskolskaya
Shubnikov
Trillat
Tutton
Wohlraube

See also books in List IV.21, Miscellaneous.

APPENDIX-V

This list contains books announced for early 1965, and books announced for 1964 (or earlier) about which sufficient information had not reached the present editor before the closing of the other lists in November 1964. It has not always been possible to verify whether the dates announced have in fact proved correct or will prove correct. For some of the books, information about their relevance to Crystallography is still lacking; those which are believed certainly relevant are marked with an asterisk, and the others must be regarded as provisional.

Author	Title, publisher, date	No. of pages	Review ref.	Subject ref.	Teaching level
l'tshuler,S.A. & Kozyrev,B.M.	(tr. from Russian) Electron paramagnetic resonance -New York: Academic Press, 1964	369		19	
melinckx,S.	The direct observation of dislocations [Supplement 6 to Solid state physics] -New York: Academic Press, 1964	487		11	
mphlett,C.B.	Inorganic ion-exchangers -Amsterdam: Elsevier, 1964	136		7	
ell,A. & Fletcher,T.J.	Symmetry groups -Nelson, Lancs.(England): Association of Teachers of Mathematics [address: Vine St. Chambers, Nelson, Lancs., England], 1964	19		2	E
erman,R. (ed.)	Physical properties of diamond -Oxford: Univ.Press, 1965	300		8	
orris,L.J. & Hauser,H.H. (eds.)	(Conference 196) Fundamental phenomena in the material sciences (Proceedings of 1st symposium on...). Vol.I: Sintering and plastic deformation -New York: Plenum Press, 1964	134		9,11	
ragg,W.L. & Claringbull,G.F. [& Taylor,W.H.]	Crystal structures of minerals [re-writing of Bragg,W.L.: Atomic structure of minerals - see List I] -London: Bell, 1965			7,15	U
ophy,J.J. & Buttrey,J.W. (eds.)	(Conference 1961, Chicago) Organic semiconductors -New York: Macmillan, 1962			7,9	
own,W.F.	Micromagnetics -New York: Interscience, 1963	143		9	
ttrell,A.H.	Theory of crystal dislocations -London: Blackie, 1964	94		11	U
xeter,H.S.M. & Moser,W.O.J.	Generators and relations for discrete groups -Berlin: Springer, (2)1964	170		6	

Appendix - V

Author	Title, publisher, date	No. of pages	Review ref.	Subject ref.	Te... i... le...
Dorfman,Y.G.	Diamagnetism and the chemical bond -London: Arnold, 1964	230		7,9	
Eitel,W.	Silicate science, Vol.II: Glasses, enamels, slags -New York: Academic Press, 1965	704		18	
Fyfe,W.S.	Geochemistry of solids: an introduction -New York: McGraw Hill, 1964	199		7,15	
Galopin,G. & Henry,N.F.M.	Introduction to the microscopic study of the opaque minerals -Cambridge (England): Heffer, 1965	200		15,14	
Goryunova,N.A.	The chemistry of diamond-like semiconductors -London: Chapman & Hall, 1965	222		8,7	
Hartmann,H.	Die chemische Bindung -Berlin: Springer, 1964	109		7	
Hyman,H.H. (ed.)	(Conference 1963: Argonne, Ill.) Noble-gas compounds -Chicago: Univ.Press, 1963	404		7	
* Ketelaar,A.A.	(tr. Becker,H.; rev.) Chemische Konstitution -Braunschweig: Vieweg, 1964	429		7	
Kripyakevich,P.I.	(tr. from Russian) A systematic classification of types of intermetallic structures -New York: Consultants Bureau, 1964	37		12	
* Krivoglaz,M.A. & Smirnov,A.A.	(tr. from Russian; ed. Chalmers,B.) Theory of order-disorder in alloys -London: Macdonald, 1964	427		11,12	
Low,W.	Paramagnetic resonance in solids [Solid state physics Suppt. 2] -New York: Academic Press, 1960	212		19,9	
*** MacGillavry,C.H.	Symmetry aspects of M.C.Escher's periodic drawings -Utrecht: Oosthoek, 1965 (for the International Union of Crystallography - see note at end of this List)	84		2,21	
McMillan,P.W.	Glass ceramics -New York: Academic Press, 1964	230		18	
Modern materials [Serial]	Vol. 3 Ed. Hausner,H. -New York: Academic Press, 1962	475		8,9,18	
Moll,J.L.	Physics of semiconductors -New York: McGraw Hill, 1964	293		9	
Moody,G.J. & Thomas,J.D.R.	Noble gases and their compounds -Oxford: Pergamon, 1964	62		7	
Pitcher,W.S. (ed.)	(Conference 1964, Liverpool) Controls of metamorphic crystallisation -Liverpool: Univ.Press, 1965				

Appendix - V

Author	Title, publisher, date	No. of pages	Review ref.	Subject ref.	Teaching level
Rostoker,W. & Dvorak,J.R.	Interpretation of metallographic structures -New York: Academic Press, 1964	230		12,13	
Salli,I.V.	(tr. from Russian) Structure formation in alloys -New York: Consultants Bureau, 1964	175		12	
Samsonov,G.V. (ed.)	(tr. from Russian; ed. Gurr,G.E. & Parker,D.J.) Refractory transition metal compounds: high-temperature cermets -New York: Academic Press, 1964	220		8,7	
Setlow,R.B. & Pollard,E.C.	Molecular biophysics -Reading (Mass.): Addison-Wesley, 1962	545		17	
Slater,J.C.	Quantum theory of molecules and solids, Vol.II: Symmetry and energy bands in crystals -New York: McGraw Hill, 1965	700		9	
Suchet,J.P.	(tr. Heasell,E., from French) Chemical physics of semiconductors -New York: Van Nostrand, 1964			7,9	
Tamas,F. (ed.)	(Conference 1963, Budapest) Silicate industry (Proceedings of the 7th conference on...) -Budapest: Akademiai Kiado, 1964			8,15	
Ubbelohde,A.R.	Melting and crystal structure -Oxford: Univ.Press, 1965	384		16,18	
Wert,C.A. & Thomson,R.M.	Physics of solids -New York: McGraw Hill, 1964	436		9	
Wilson,A.J.C.	(tr. from English of 1963) Rontgenstrahl-Pulverdiffraktometrie: mathematische Theorie -Eindhoven: Centrex, 1965	152		4,5	
Winchell,A.N. & Winchell,H.	Microscopical characters of artificial inorganic solid substances or artificial minerals [see Winchell,A.N., List I] -New York: Academic Press, (3)rev.1964	350		T,7,15	
Winchell,H.	Optical properties of minerals -New York: Academic Press, 1964	100		15	
[Ždanov,G.S.=] Zhdanov,G.S.	Crystal physics -Edinburgh: Oliver & Boyd, 1965			9	

*** Note: "Symmetry aspects of M.C.Escher's periodic drawings", by C.H.MacGillavry, was specially commissioned by the International Union of Crystallography as a help to the teaching of crystallography. There are 12 coloured illustrations, and 29 black-and-white.

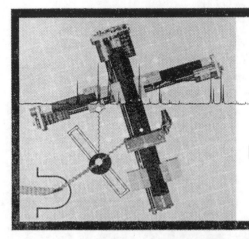

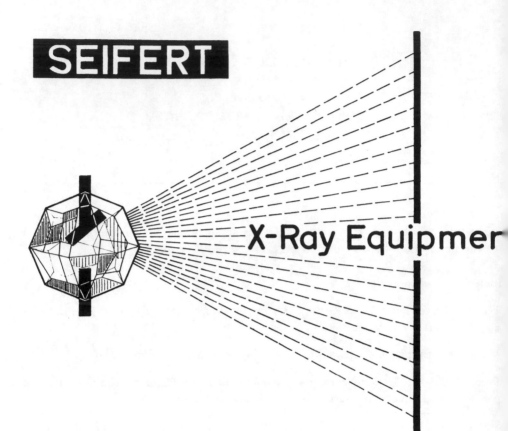

INTERNATIONAL UNION OF CRYSTALLOGRAPHY

List of Publications

ACTA CRYSTALLOGRAPHICA

Acta Crystallographica is a scientific journal published in monthly issues and containing original articles in English, Frenc and German dealing with new crystal structures, refinements of known structures, new theoretical and experimental methods o structure determination, the theory of diffraction, computing methods, apparatus, and various other related topics.
Subscription price per annual volume: 240 Danish Kroner, post free ($35 or £12.10s. at the present rates of exchange).
For particulars about prices of back numbers, and preferential subscription rates for *bona fide* crystallographers, please wri to the publishers, Messrs Munksgaard, Prags Boulevard 47, Copenhagen S, Denmark.

STRUCTURE REPORTS

Vols. 8–20, covering the years 1940–1956, have appeared. Further volumes are in preparation. Vol. 14 is a supplementa volume and cumulative index for 1940–50.

Volume	8	9	10	11	12	13	14	15	16	17	18	19	20
Years covered	1940/41	1942/44	1945/46	1947/48	1949	1950		1951	1952	1953	1954	1955	195
Pages	384	448	325	779	478	644	215	588	651	863	846	692	728
Price in Dutch florins	80.—	70.—	55.—	100.—	70.—	100.—	35.—	110.—	120.—	125.—	120.—	100.—	100.-
* Price in U.S. dollars	22.50	19.50	15.50	28.—	19.50	28.—	10.—	31.—	33.50	35.—	33.50	28.—	28.-

For prospectus, order form, and particulars about preferential prices for *bona fide* cystallographers, please write to th publishers, N.V. A. Oosthoek's Uitgevers Mij, Domstraat 11-13, Utrecht, The Netherlands.

INTERNATIONAL TABLES FOR X-RAY CRYSTALLOGRAPHY

This successor to *Internationale Tabellen zur Bestimmung von Kristallstrukturen* has been published in three large-size volum and the publication of additional volumes is being considered.
Vol. 1. **Symmetry Groups.** Published 1952, pp. xii + 558, price £5.5s.
Vol. 2. **Mathematical Tables.** Published 1959, pp. xviii + 444, price £5.15s.
Vol. 3. **Physical and Chemical Tables.** Published 1962, pp. xvi + 362, price £5.15s.
For prospectus, order form. and particulars about preferential prices for *bona fide* crystallographers, please write to t publishers, The Kynoch Press, Witton, Birmingham 6, England.

FIFTY YEARS OF X-RAY DIFFRACTION

This commemorative volume contains the history of the early discoveries, a survey of the development in the various fiel of X-ray crystallography, and of Schools and Research in the various countries, and Personal Reminiscences of some thi crystallographers. x + 733 pp.; price 40 Netherlands Guilders or 11.25 U.S. dollars *.
Publishers: N.V. A. Oosthoek's Uitgevers Mij, Domstraat 11-13, Utrecht, The Netherlands.

WORLD DIRECTORY OF CRYSTALLOGRAPHERS

The second edition of this compilation appeared in 1960, and contains biographical information concerning 3557 crystallograph and other scientists from 54 countries. Price 1.50 U.S. dollars.
A third edition is in preparation.
Publishers: N.V. A. Oosthoek's Uitgevers Mij, Domstraat 11-13, Utrecht, The Netherlands.

* The dollar prices are subject to changes in the official rates of exchange without prior notice.

The publications can also be ordered from Polycrystal Book Service, G.P.O. Box 620, Brooklyn 1, N.Y., U.S.A.; or from any booksel

---